Plastic and Polymer Industry by Region

Ololade Olatunji

Plastic and Polymer Industry by Region

Production, Consumption and Waste
Management in the African Continent

 Springer

Ololade Olatunji
Research and Product Development
Geo Calibrations
Lagos, Nigeria

ISBN 978-981-19-5233-3   ISBN 978-981-19-5231-9 (eBook)
https://doi.org/10.1007/978-981-19-5231-9

This Springer imprint is published by the registered company Springer Nature Singapore Pte Ltd.
The registered company address is: 152 Beach Road, #21-01/04 Gateway East, Singapore 189721,
Singapore

Preface

One of the most commonly used classes of polymers, plastic, has come to the world's attention for the adverse environmental impact posed by these materials. Although seemingly irreplaceable in some applications, some of these applications can be met with other more sustainable polymers. With the need to adopt a more sustainable circular economy, the future of the plastic and polymer industry lies in re-evaluating the early developments, the current trends in production and consumption, and the management of the waste from this industry. These issues vary in different parts of the world and therefore each region should be considered separately.

While Africa is made up of several different countries which might be different from each other in many ways, these countries within the African continent have some commonalities such as physical region, some shared history, resources, and some shared policies through organizations such as the African Union, African Free trade Zone, and ECOWAS. With a population of over a billion, the African continent has become an attractive market for various businesses. Several publications in recent years have pushed for the advancement of the African continent towards increased manufacturing as a road to development. This inevitably includes the plastics and other polymer industries. Careful consideration must be taken to ensure that this growth will focus on more sustainable and greener manufacturing; otherwise, this anticipated growth in the plastics and polymer industry will only spell increased pollution and worsening of the environment.

These facts have motivated this book on plastics and polymers in the African region. The book seeks to address the need for a technically guided thought on the production, consumption, and waste management of plastic and polymers, focusing on the African continent. Issues such as resource availability, processing technologies, plastic policies, and much more are covered in the book. The book also includes original images from some cities and towns in parts of Africa to provide pictorial examples for enhanced reader experience.

Lagos, Nigeria   Ololade Olatunji

Contents

Chapter 1
Introduction

Polymers have come to be recognized as the most important materials of this age. From the polymers that have existed since the beginning of life to those that have recently been synthesized. Plastics, the most widely used form of polymers in the modern age, have brought much attention to polymers in recent years. As of 2014, of the total global single-use plastic production, Africa produces only 1%. North America produces 21%, Europe 16%, northeast Asia 26%, the former USSR 3%, Asia and the Pacific 12%, Central and South America 4% and the Middle East produces 17%. These are single-use plastics made of LDPE, HDPE, PS, and EPS excluding PET and PP (UNEP, 2018). However, the plastic waste generation from Eastern, Western, Northern, and Middle Africa is in the top ten when plastic waste littering is compared among regions (Lebreton & Andrady, 2019). When plastic waste littering is compared by country, three African countries make the top ten. Nigeria comes sixth generating an estimated 1.9 million tonnes per year, Tanzania comes seventh with 1.77 million tonnes and Egypt is tenth with 1.6 million tonnes of plastic waste litter generated annually.

So far, more plastic bans have been implemented across Africa compared to any other region (UNEP, 2018). The Nile and Niger are among the top 10 rivers through which 90% of plastic waste flow into the world's oceans. The Nile contributes 84,792 tonnes while the river Niger contributes 35,196 tonnes annually (Schmidt et al., 2017; UNEP, 2018). These data suggest interesting dynamics in the production and consumption of plastics within the African continent.

As many countries across Africa become more urbanized, consumption of more modern materials increases. Polymers have become the most ubiquitous materials of the modern age that are present in almost every aspect of modern life. To guide the development of such an important industry towards a more sustainable one, the issues around the production, consumption, and management of waste from the industry must be evaluated from diverse perspectives. This book takes the approach of looking at the polymer industry with the focus on a specific region since these issues vary significantly from one region to the other.

O. Olatunji, *Plastic and Polymer Industry by Region*,
https://doi.org/10.1007/978-981-19-5231-9_1

The next chapter of this book, Chap. 2, goes back a few centuries to what polymers existed in the pre-industrial age in Africa and the trade and industry that occurred in relation to polymeric materials like rubber and cotton. Some of the processing techniques used to extract and process some of these polymers are outlined in Chap. 3. This gives an idea of the range of polymers available in the pre-industrial era and how they were processed. Chapter 4 then looks at the plastics and polymers and the rate of consumption in Africa today. Some case studies and data on plastic and polymer consumption are presented. Chapter 5 covers the topic of plastic bans across Africa, the approaches are taken, and the impact. Plastic recycling in Africa is covered in Chap. 6. The chapter looks at the recycling activities in the formal and informal sectors and discusses the local and international organizations involved in recycling activities.

Chapter 7 looks at biodegradable plastics and polymers, discussing the current state, and the opportunities and the potential for biopolymers in the continent. In Chap. 8, the plastic and polymer manufacturing within the African region is explored by looking at the type of processes and some existing plastic manufacturers. Chapter 9 looks at chemical plastic recycling and energy recovery. The chapter begins with a general overview and then looks at the existing, emerging, and potential opportunities within Africa. Chapter 10 discusses the issue of plastic waste management in Africa. Some specific countries and cities are discussed with examples. The issue of microplastics is discussed in Chapter 11. The contribution of the African region to microplastics in the world's oceans, the impact, and current measures to address microplastic wastes among other issues is discussed. In Chap. 12, a future outlook for the plastic and polymer industry in Africa is discussed, sharing insights from some of the topics reviewed in other chapters. The book then ends with conclusions and recommendations in Chap. 13.

References

Lebraton, L., & Andrady, A. (2019). Future scenarios of global plastic waste generation and disposal. *Palgrave Communications, 5*(6), 1–11.
Schmidt, C., Krauth, T., & Wagner, S. (2017). Export of plastic debris by rivers into the sea. *Environmental Science and Technology, 51*(21), 12246–12253.
UNEP. (2018). *Single use plastics: A roadmap for sustainability.*

Chapter 2
Polymers in Early Trade and Industry in Africa

Abstract This chapter begins with an overview of the origins of plastics and polymers and the need cases that motivated the invention of plastics. It then goes on to analyze the existing industry before the introduction of synthetic plastics and polymers in Africa. We look at some of the major polymer-based materials that were produced and traded in the early African market. This informs the range of polymer-based materials that were available during the period before the introduction of synthetic and crude-oil-based plastics and polymers.

Keywords Polymers · Industry · Sisal · Rubber · Textiles · Cotton

Origins of Plastics and Polymers

Plastics have become an essential part of our existence as modern humans. When compared to other materials like glass, ivory, and metals which have been in use for thousands of years, plastics have not been around that long. The scientific understanding of the macromolecular nature of plastics and polymers did not exist until the 1920s thanks to Staudinger (Billmeyer, 1981). The first known plastic invention was derived from reacting cellulose with camphor to obtain celluloid. There was a need to meet the demand for ivory used to make billiard balls without further endangering the elephants from which ivory was obtained. However, the first truly synthetic polymer made without any material input from nature is Bakelite. The invention of Bakelite was motivated by the need for the replacement of the naturally occurring material shellac, which found use in electrical insulations.

The discovery of celluloid and then Bakelite was followed by the discovery of other plastics such as Nylon. With the advancement in the understanding of the polymerization process, more plastics emerged. World War II saw the applications of plastic extended to many areas such as bulletproofs and water-resistant casings and lightweight containers. After the world war with increased spending, the use of plastics increased and continued to increase into the twenty-first century. The most common plastics with resin identification numbers 1 to 6 which are in commercial use today were introduced into the market between 1936 and 1984. PET was introduced in 1944, PVC in 1936, PE in 1955, PP in 1957, and PS in 1937 (Billmeyer, 1981).

Staudinger later went on to win the Nobel Prize in 1953 for his work on the theory of macromolecules (which included plastics). Today plastic gets used in everything from healthcare to construction to food packaging and more. They have brought significant improvement in the quality of life. This includes clean and convenient drinking water in lightweight portable bottles, insulated parts for electronics, prosthetics and scaffolds in tissue engineering, breathable clothing, and so much more.

The Early Trade and Industry in Africa

Within the scope of this text, the early industry in Africa refers to the period between the 1920s and 1960. Before this period, the periods before the 1900s are referred to as the pre-industrial period in Africa within the scope of this chapter. Between the 1700 and 1800s, the chartered companies of Europe dominated trade in Africa. These companies included the Royal African Company, the Dutch Western India company, the Portuguese Guinea Company, the British South Africa Company, and the Royal Niger Company (Ringrose, 2001). The industrial period can be further divided into the period during the colonial regime between the 1920s and the 1940s and the post-colonial industrial period between the 1950s and the 1960s after much of the colonies had just gained independence (Mendes et al., 2014). The period between the 1960s and the 1980s marks the early industry period where many African countries adopted the structural adjustment program introduced by the IMF and the World Bank.

Africa had been importing European-manufactured goods since the 1500s when parts of Africa like Sao Tome Principles had been occupied by Europeans. Benin, Oyo, and Kongo traded ivory alongside gold and slaves between the 1600 and 1800s. They exchanged these for textiles as well as weapons and other goods from Europe. While there is a tremendous amount of historically significant events and details in this period of history, the focus of this book is the polymer resources that were produced, consumed, and traded during this era.

Between the 1920s and the 1960s, many African countries adopted import substitution. This was intended to boost the industrialization of the countries. However, as some authors noted, often this resulted in further importation of more goods. This includes machinery to produce the goods domestically and additional parts and materials that were not available locally (Hope & Misir, 1981). Small-scale production of products that required mostly local raw materials like textiles, bottles, soaps, and cigarettes began in the early 1920s in Kenya, Zimbabwe, and Congo Brazzaville. Later followed the production of plastics (Mendes et al., 2014). Polymer-based products such as textiles, rubber, and plastics were part of the products being manufactured in the early industries in Africa. Much of the products of the early industry were those which could be made with more local raw materials and less imported inputs. It is thus expected that much of the polymer products would have been those based on raw materials such as cotton, rubber, grains, and starch. End products such as clothing, shoes, tires, and alcohol were largely imported. Table 2.1 summarizes the timelines in the trade and industry of polymer-based materials in Africa between

Table 2.1 A timeline of polymer-based materials trade and industry in Africa between 1400 and 1980s

Before 1400	Trading of natural polymers such as those sourced from animal and plant matter for applications such as food, shelter, ornaments, medicines, and clothing
1400–1500	Trading of ivory alongside other commodities
1890	Rubber trade
1920–1940	Colonial-led industry. Production of textiles, paper, plastics products, and plastic components in products like cigarettes
1950s–1960s	Import substitution industry strategy implemented in Nigeria, Tanzania, and Zambia
The 1960s	Post-independence industry
The early 1980s	Ghana and Madagascar and other sub-Saharan African countries adopt the import substitution strategy
Mid-1980s	Structural adjustment program was implemented in many African countries

the 1400s and the late 1980s. This is based on information extracted from different literature referenced within this chapter.

Polymers of the Early Trade and Industries in Africa

This section will look at some polymer-based goods that were traded and produced from the 1500s to the 1900s. This section will identify the polymer composition of these products and present some data on how they were traded and produced.

Silk

Silk is a proteinous fiber that is produced by silkworms. The core of silk is the silk fibroin and the surrounding outer layer is the sericin. The sericin and fibroin are then covered in a lipid layer that holds the fiber together (Costa et al., 2018). Silk has been produced in parts of Africa such as Madagascar and Nigeria since pre-industrial times. For example, the Yoruba people of Nigeria produced silk known in local parlance as "*Sanyan*". These were produced from the Anaphe moth (McKinney & Eicher, 2009). Reports from as far back as 1908 indicate that during this period in Nigeria, silk was produced in Ibadan, Ilorin, Bauchi, Bornu, and other regions. The different types of silk were also being traded from one region to the other. For example, Dudgeon (1908) reported that a pure white type of silk called *Gambari* or *Hausa silk* was brought to markets in Ibadan in the western part of Nigeria from Bauchi and Bornu in the northern part. This *Sanyan* silk was originally used for the production of the highly priced fabric called "*Aso oke*". Today, silk fiber has been

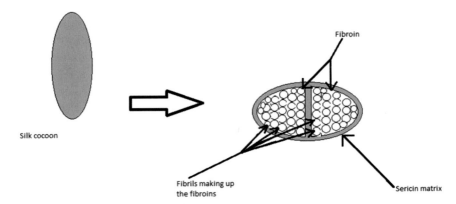

Fig. 2.1 Illustrated structure of silk cocoon and cross section if silk fiber showing fibroin and sericin

largely replaced by cheaper substitutes and synthetic yarns in the production of aso oke.

The production of silk in Africa dates even further back to the 1700s. Attempts at large-scale production of silk from the silkworm bombyx mori were in 1726 by the Dutch East India Company in Southern Africa (Scholtz & Snyman, 1982). When nylon was invented, it became a cheaper substitute for silk (Mckinney & Eicher, 2009). This no doubt had an impact on the demand for silk. Figure 2.1 shows an illustration of silk cocoon, sericin, and fibroin based on information from different texts (Naomi et al., 2020; DeBari & Abbott, 2018; Patel and Singh, 2020).

Rubber

Natural rubber comprises polymer polyisoprene. It is obtained from the rubber trees. Hevea Brasiliensis is the main commercial source of high-yield rubber containing mainly cis 1,4-polyisoprene. Other strains include Balata and Gutta-percha which are made up of mostly trans 1,4-polyisoprene (Joseph & Ebdon, 2014). Some publications suggest that rubber was exported from Lagos, Nigeria in 1895 (Omosini, 1979). Other texts also reported that rubber was shipped widely from East African regions including Tanzania from the 1890s (Monson, 1990). Much of the rubber collected at this time grew wild. The rubber was then brought to coastal regions like Lagos in Nigeria and Zanzibar in Tanzania to be shipped to Europe. According to the UNSTATS data, Nigeria and Cote d'Ivoire are leading African countries in rubber production in recent years. Table 2.2 shows figures for rubber production in Nigeria, Cote d'Ivoire, and the entire African continent in 1970 and in 2018 quantified by area harvested. The sum of area harvested in Nigeria and Cote d'Ivoire in 2018 accounted for more than half of the area harvested for the entire African continent in the same year.

Table 2.2 Area harvested for rubber in Nigeria and Cote d'Ivoire and the entire African continent in 1970 and 2018		Area harvested in 1970 (ha)	Area harvested in 2018 (ha)
	Nigeria	95,000	361,779
	Cote d'Ivoire	10,406	285,014
	Africa	338,406	928,960

Ivory

Ivory comprises mainly of magnesium-rich carbonated hydroxyapatite mineralized collagen (Alberic et al., 2018). These are obtained from the elephant tusks. Ivory is valued for its fine grain and luster and the smooth surface feels and hardness. It is mainly used in ornaments, jewelry figurines, fashion, and other high-end applications. Ivory was previously used in the production of billiard balls before the invention of plastics. Indeed the switch in the production of billiard balls from ivory to polymers reduced the pressure on the ivory resource. It is even said that plastics in a way saved the elephants from extinction.

Sisal

The sisal plant belongs to the agave plant family. The main producers of this cellulosic fiber material are Tanzania, Kenya, Madagascar, and Brazil. It is grown in much fewer quantities in South Africa, Mozambique, Mexico, Haiti, and China (FAO, 2017). Sisal produced in Africa is of higher value than that produced in Brazil. This is mainly due to the difference in the production process. The production processes are covered in another chapter in the book. The most common use for sisal is in the production of woven products like baskets, handbags, hats, and mats. UN data on sisal production dates back to 1965 (FAO, 2000). Its origin dates back to the 1800s in Mexico; however, it was introduced to Africa in the late nineteenth century (Trejo-Torres et al., 2017). In Kenya, records of sisal production date back to 1914, and sisal production showed rapid growth for 5 decades between 1914 and the mid-1960s (Phologolo et al., 2012) before the introduction of synthetic fibers. Figure 2.2 shows a sisal plant being used as ornamental plant observed at a park in Lagos Nigeria.

Cotton

A cellulosic material sourced from the cotton plant and is widely used in textile production. Cotton is made up of 88–97% cellulose while the remaining 3–12% comprises pectins, waxes, and proteins (Candido, 2021). The cellulose in cotton is highly crystalline, well ordered, and has the highest molecular weight compared to

Fig. 2.2 Sisal plants at the Johnson Tinubu Jakande Park in Lagos Nigeria where it is mainly used as an ornamental plant, May 2022

cellulose found in other plants (Hsieh, 2007). Cotton is the most abundant source of pure cellulose in plants when compared to fibers such as jute and flax.

Archeological studies show evidence of cotton in Africa dating back to between 2600 and 2400 BCE in Nubia, around the tropical Nile Valley (Kriger, 2005). In the pre-industrial era, the Sokoto Caliphate, which was said to be the most populous state in the tropical African region, exported cotton cloth to other parts of West Africa and North Africa. Other records show that more modern cotton production originated in India and from there spread to Iraq in the 1600s and into other parts of the world, reaching Africa through Tunisia and Morocco and through the Nile Valley into other parts of Africa (Pomeranz & Topik, 1999). Global cotton production has increased significantly from 13.8 million in 1980s to 27 million tonnes in 2018 (Negm & Sanad, 2020). Cotton has always played an important role in trade in Africa and across the world.

Leather

The main component of leather is collagen which is a polymer material present in the connective tissues of animals. Leather is obtained from the processing of animal hides to eliminate other compounds and strengthen the collagen material. One report suggests that leather was being produced in Northern Nigeria in the 1900s (Adebayo, 2009). By the 1950s Madagascar's central province Imerina was producing leather goods on a large scale (Campbell, 1991). The leather trade is said to originate from the Northern parts of Nigeria in 1887 (Sani, 2009) in the form of goat skins that had been tanned and dyed. These were also made into other commodities like sandals. These were then traded and exported to other parts of the region and the world.

Casava

Casava contains between 8 and 28% amylose starch and up to 90% carbohydrate (Wheatley & Zakhia, 2003), and is today a staple in many African countries as a drought-resistant tuber. Its applications include food, beverages, animal feed, and other uses. Cassava is not native to Africa. Earlier thought to originate from South America, later studies suggest that it originates from the South of the Amazon Basin (Olsen & Schaal, 1999). It was later introduced to other parts of the world. In Uganda, for example, cassava was introduced around 1862 to 1875 by Arab traders. Its production has since then reached millions of tonnes.

Gum Arabic

This is a naturally occurring polymer obtained as an exudate from the stem and branches of the acacia tree. It grows most commonly in Sudan, it is also abundant in other African countries including Ethiopia, Nigeria, Somalia, and Senegal. However, it is most abundant in what is referred to as the African Gum Belt. Gum Arabic is said to have been used from as far back as the time of the ancient Egyptians when they used them for pigments. In recent years, gum Arabic makes up between 10 and 20% of Sudan's total national income. Today, applications of gum Arabic cut across several industries including food, medicine cosmetics, pharmaceutics, textiles, and beverages (Ahmed et al., 2018).

The production quantities in tonnes of these polymer-containing materials for different countries in Africa are listed in Table 2.3. The year 1961 was chosen as it was around this period that many African countries gained independence from colonization. Different countries were selected to give an idea of the varying production countries across the continent for different products; however, these countries listed are not the only producers.

Challenges Faced by the Early Polymer Trade and Industries of Africa

The early industries faced the challenge of lack of human capital, lack of technology, and poor administration. This led to a high cost of production because equipment

Table 2.3 Production quantities of some polymer raw materials in different countries 1961 (UN statistics)

Polymer raw material	Country	Production quantity (tonnes)
Rubber	The central African Republic	531
Sisal	Angola	58,800
Cassava	Benin	280,000
Silk	Madagascar	100
Cotton lint	Benin	1,400
Cottonseed	Nigeria	155,000
Jute	Zimbabwe	1,000
Maize	Uganda	196,000
Flax fiber and tow	Egypt	9617
Jute	Egypt	5000

and inputs had to be imported. Inadequate facilities and machinery as most of the plants and production units were obsolete. Production efficiency was also poor. Since the technology and machinery had to be imported, the cost of production of many products was high as a result.

Away from the industrialized systems, nationals engaged in small-scale agriculture, trading, and crafts. Cellulosic fibers such as sisal were hand woven into craft products such as baskets and textiles. As local craftsmen had to compete with cheaper imports, they began to adjust their business activities to remain competitive. For example, the textile makers would reduce the thickness of the material to use fewer materials. Some reports state that the handcraft businesses were already being affected by the imports even before the 1920s (Austin et al., 2017). Some of these handcraft production were said to survive the competition due to factors such as the high quality of African iron in iron smelting and the intricate high-end woven fabrics like Kente from Ghana and Machila cloth from Malawi. These and other handcrafted products were not easily replaced by imports.

Conclusion

A wide range of polymeric raw materials is produced in different countries across the African continent. Various reports show that the production and trade of polymeric materials such as cotton have existed from as far back as 2600 BCE. Therefore, although the chemistry of the synthetic polymers we know today was not discovered until the early twentieth century, polymers have been in use long before the term polymer was coined. Some of the polymers produced in Africa originated from other parts of the world through trade. Polyisoprenes, cellulose, collagen, and starch are some of the polymers that comprise the polymeric raw materials covered in this chapter. The early polymer industry involved mainly trading these raw materials in exchange for other commodities or money as well as processing these materials into goods such as clothing, food, and tools. The production and processing techniques, as well as applications of these polymers, will be discussed in the next chapter.

References

Adebayo, A. G. (2009). The production and export of hides and skins in colonial Northern Nigeria, 1900–1945. *The Journal of African History., 33*(2), 273–300.

Ahmed, R. H. A., Mariod, A. A., & El Sanusi, S. M. (2018). Effect of gum Arabic on the fecal bacterial mass in healthy human volunteers. In *Gum Arabic: Structure properties, applications, and economics* (pp. 297–304).

Alberic, M., Gourrier, A., Wagermaier, W., Fratzl, P., & Reiche, I. (2018). The three-dimensional arrangement of the mineralized collagen fibers in elephant ivory and its relation to mechanical and optical properties. *Acta Biomateriala, 72*, 342–351.

Austin, et al. (2017). Patterns of manufacturing growth in sub-Saharan Africa: From colonial to present. In O'Rourke & Williamson (Ed.), *Spread of modern industry to the periphery since 1871* (p. 345). Oxford University Press. ISBN: 9780198753643.

Billmeyer, 1981Billmeyer Jr., F. W. (1981). Textbook of polymer science. Wiley-Interscience. ISBN: 0-471-03196-8.

Campbell, G. (1991). An industrial experiment in pre-colonial Africa: The case of imperial Madagascar, 1825–1861. *Journal of Southern African Studies, 17*(3), 525–559.

Candido, R. G. (2021). Recycling of textiles and its economic aspects. In *Fundamentals of natural fibres and textiles* (pp. 599–624). The Textiles Institute Book Series.

Costa, F., Silva, R., & Boccaccini, A. R. (2018). Fibrous protein-based biomaterials (silk, keratin, elastin and resilin proteins) for tissue regeneration and repair. In *Peptides and proteins as biomaterials for tissue regeneration and repair* (pp. 175–204).

DeBari, M. K., & Abbott, R. D. (2018). Microscopic considerations for optimizing silk biomaterials. *Wires Nanomedicine and Nanobiotechnology, 11*(2), e1534.

Dudgeon, G. C. (1908). Silk-producing insects of West Africa. *Nature Publishing Group, 79*(2041), 160.

FAO Committee on Commodity Problems. (2017). *Review of the sisal market industry: prospects and policy.* CCP: HFJU 17/2.

Hope, K. R., & Misir, D. (1981). Import substitution strategies in developing countries—A critical appraisal with reference to Guyana. *Foreign Trade Review, 16*(2), 153–165.

Hsieh, Y. L. (2007). Chemical structures and properties of cotton. In S. Gordon & Y. L. Hsieh (Eds.) *Cotton science and technology* (pp. 3–34). Woodhead Publishing Series in Textiles. ISBN: 9781845690267.

Joseph, P., & Ebdon, J. R. (2014). Recent developments in flame retarding thermoplastics and thermosets. In A. R. Horrocks & D. Price (Eds.), *Fire retardant materials* (pp. 220–263). Woodhead Publishing. ISBN: 9781855734197.

Kriger, C. E. (2005). Mappin the history of cotton textile production in pre-colonial West Africa. *African Economic History, 33*, 87–116.

McKinney, E., & Eicher, J. B. (2009). Unexpected luxury: Wild silk textile production among the Yoruba of Nigeria. *Textile: Cloth and Culture, 7*(1), 40–55.

Monson, J. (1990). From commerce to colonization: A history of the rubber trade in the Kilombero Valley of Tanzania, 1890–1914. *African Economic History, 21*, 1130130.

Naomi, R., Ratanavaraporn, J., & Fauzi, B. M. (2020). Comprehensive review of hybrid collagen and silk fibroin for cutaneous wound healing. *Materials, 13*, 3097.

Negm, M., & Sanad, S. (2020). Cotton fibres, picking, ginning, spinning and weaving. *Handbook of natural fibres. Volume 2: Processing and applications* (2nd ed., pp. 3–48). The Textile Institute Book Series.

Olsen, K. M., & Schaal, B. A. (1999). Evidence on the origin of cassava: Phylogeography of Manihot esculenta. *PNAS Proceedings of National Academy of Sciences of the United States of America, 96*(10), 5586–5591.

Omosini, O. (1979). The rubber export trade in Ibadan, 1893–1904: Colonial innovation or rubber economy. *Journal of Historical Society of Nigeria, 10*(1), 21–46.

Patel, M., & Singh, P. S. (2020). Phenomenological models of Bombyx mori silk fibroin and their mechanical behavior using molecular dynamics simulations. *Materials Science and Engineering C, 108*, 110414.

Phologolo, T., Yu, C., Mwasiagi, J. I., Muya, N., & Li, Z. F. (2012). Production and characterization of Kenyan sisal. *Asian Journal of Textile, 2*(2), 17–25.

Pomeranz, K., & Topik, S. (1999). In M. E. Sharpe (Ed.), *The world that trade created: Society, culture and the world economy, 1400 to the present*. Armonk. New York. ISBN: 0-756-0250-4.

Ringrose, D. R. (2001). *Expansion and global interaction 1200–1700*. Longman.

Sani, A. B. (2009). Raw material production for export in Northern Nigeria: The experience of the people in the livestock and allied industries under British rule C. 100–1960. *African Economic History, 37*, 103–127.

Scholtz, C. H., & Snyman, A. (1982). Results of a detailed investigation into commercial silkworm (Bombyx Mori L.) rearing in southern Africa. *Phytophylactica, 14*, 119–122.

Trejo-Torres, J. C., Gann, G. D., & Christenhusz, M. J. M. (2017). The Yucatan Peninsula is the place of origin of sisal (Agave sisalana. Asparagaceae): Historical accounts, phytogeography, and current population. *Botanical Sciences—Taxonomy and Floristics 96*(2), 366–379.

Wheatley, C. C., & Zakhia, G. C. N. (2003). Cassava∣ the nature of the tuber. In B. Caballero (Ed.), *Encyclopedia of food sciences and Nutrition* (2nd ed.). Academic Press. ISBN: 9780122270550.

Chapter 3
Production and Processing Techniques of Some Polymeric Materials in Early Trade and Industry of Africa

Abstract This chapter explores the different techniques that were used in the early industry of Africa in the processing of polymer-based products and raw materials. Some of these techniques are still in use today while some are long gone. The motivation here is to review the pre-existing methods for producing certain products and understand the need for advancement or transition to other methods. In some cases, we are able to draw some connections to existing polymer processing technologies and readers may also be able to gain some ideas on alternative advancements perhaps towards more sustainable advanced methods. The following sections discuss the processing methods of some of the polymer-based products from Chap. 2.

Keywords Silk · Bombyx mori · Collagen · Nanocellulose · Spinning · Weaving

Production and Processing of Silk into Textiles

Silk is most widely used in the production of textiles for garments and accessories. Other uses include filler fiber to improve the feel of lightweight fabrics and anti-allergic beddings. Silk sericin, a by-product of silk production and silk fibroin, today has found applications in biomedical engineering and drug delivery (Lamboni et al., 2015). The waste from silk production also finds use in papermaking and as a fusing agent (Slegtenhorst & Venter, 2009).

Like other parts of the world, in Africa, silk is regarded as a superior material to other fibers such as cotton and wool. In different parts of Africa, silk fabrics are regarded as prestigious and used for many important purposes. For example, in Madagascar, Borocera Madagascariensis silk is used in the production of textile for burial shrouds, and in Nigeria, the Yoruba people use silk from the Anaphe moth known as *Sanyan* in a ceremonial outfit called *aso oke*. In the unprocessed form, the San people of Namibia use the silk cocoon of Gonometa Postica to make ankle rattles (Slegtenhorst & Venter, 2009). As mass-manufactured fabrics dominate the textile industries, the silk production in Africa today survives on the cultural significance some of the silk garments hold.

Table 3.1 Various silk types are found in different parts of the African continent

Silk type	Source	Region/country	Reference
Kalahari wild silk	Gonometa Postica	Southern Africa	Slegtenhorst and Venter (2009)
Landibe	Borocera madagascariensis	Madagascar	Spring and Hudson (2002)
Sanyan	Anaphe Moth	Nigeria, Kenya, Uganda, Mozambique	McKinney and Eicher (2009)
Epiphora silk	Epiphora bauhiniae	Sudan and West Africa	Piegler (2020)

Types of Silk

Although there are at least eight known types of silk, today the most commonly used silk for the production of garments is sourced from the Bombyx mori moth. It is called mulberry silk and is cultivated in China. Some other types of silk found in Africa are listed in Table 3.1. Of these, the Anaphase silk is produced in Nigeria, Kenya, Uganda, and Mozambique while the Borocera madagascariensis is produced in Madagascar (Spring and Hudson, 2002). Recent reports also indicate wild silks from the Saturniidae, Notodontidae, and Lasiocampidae moths exist in Asia and Africa. According to their entomological origins, they are the Kalahari wild silk produced in Southern Africa, *landibe* silk in Madagascar, Sanyan silk in west Africa, and Epiphora silk in Sudan and West Africa (Peigler, 2020).

Processing of Silk

Today, the technology for processing silk cocoons into fiber include electrospinning, 3D printing, and other advanced methods to achieve high-quality silk fabrics (Reddy, 2020). Traditional processing of silk in Africa comprises the main stages of degumming, spinning of the yarn, dyeing, and weaving of the narrow strips. Prior to the processing, the silk must be collected from the source. Figure 3.1 is a schematic diagram summarizing the silk production process.

Silk Cocoon Formation

In the early industries, silk cocoons were obtained from the wild. The cocoons of the moths growing on trees were collected for processing. However, the yield from the wild-growing silk sources were below that is needed to supply a global market. Much of the investigation into the manufacturing of silk in Africa has been directed towards the cultivation of the Bombyx mori silkworm in parts of Africa like Nigeria.

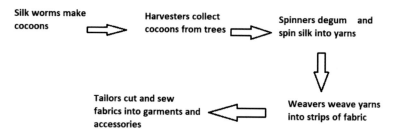

Fig. 3.1 Schematic diagram of silk production process

This Bombyx mori silkworm is endemic to China and its mass cultivation in African countries proved difficult.

The Anaphe silk worm of the genus Anaphe infracta, Anaphe venata, and Anaphe moloney are the most prominent in silk production in parts of Africa like Nigeria (McKinney and Eiccher, 2009). Although there are several other types, the silkworms are known to live in groups always close together and at night they tend to gather on top of each other to sleep. They spin their silk cocoons also in groups. They gather together within a safe pocket of their tree of choice. The cocoons serve as a protective nest for the silkworms to develop their wings and metamorphosis into moths. The colors of the cocoons vary for different types of silkworms and environments. It generally ranges from whitish to reddish-brown. To form the cocoon the silk spun by the silkworms is held together by a gummy secretion, the sericin, assuming a composite structure where the silk fibroins serve as the reinforcement within the matrix formed by the gummy secretion. The cocoon has a fibrous outer layer and an inner hard shell although the exact thickness, size, and structure vary among species and in a different environment. The properties of silkworm fibers vary with different types of silkworm species. Bombyx Mori, for example, produces colors of different types from white to yellow–red or green (Wang et al., 2014). The protein analysis has also shown that the amino acid composition of silk from different insects varies. For example, while Bombyx mori has a glycine content of 43.45 mol/g that of Anaphe is 27.65 mol/g (Kebede et al., 2014). Table 3.2 compares some silkworm fiber properties of Bombyx mori and Anaphe.

Harvesting and Gathering of Silk Cocoons

The silk cocoons are collected from the trees on which the silkworms gather and form them. Some tree species are identified as being used by silk work for nesting. For example, the Gonometa postica forms its cocoons on acacia trees while the caterpillars form their silk cocoon on tamarind trees. The harvester can be farmers, hunters, or herdsmen who collect the cocoons as they come by the trees in their daily activities and journeys. Harvesters can also be dedicated silk cocoon collectors. These are then traded at markets, mostly to spinners.

Table 3.2 Some physical properties of silkworm fiber from Bombyx mori and Anaphe

Silkworm types	Fiber properties	Value	Reference
Bombyx mori	Color	White, Yellow–red or Green	Wang et al. (2014)
	Break stress	0.427GPa	Kebede et al. (2014)
	Young modulus	8.787GPa	
	Strain at break	21.8%	
	Fiber diameter	7.38 microns - 24.16	Chen et al. (2019)
	Luster	Better than anaphe	McKinney and Eicher (2009)
	Resistance to rot	Less than that of Anaphe	
	Glycine content	43.45 mol/g	Kebede et al. (2014)
	Valine content	25.7 mol/g	
Anaphe	Color	A shade of brown	Egelyng et al. (2017)
	Break stress	0.406GPa	Kebede et al. (2014)
	Young's modulus	8.161GPa	
	Strain at break	15.4%	
	Fiber diameter	0.0004 - 0.0007 inches	McKinney and Eicher (2009)
	Luster	Less than Bombxy Mori	
	Resistance to rot	(better than Bombyx mori)	
	Glycine content	27.65 mol/g	Kebede et al. (2014)
	Valine content	1.23 mol/g	

For silkworms produced in sericulture, the silkworms are bred on a farm or in an enclosed environment where they are allowed to spin their cocoons. In one method, the silkworms are enclosed in a closed calabash. The darkness is thought to result in the production of white cocoons. Wild silkworms have coarser textures than cultivated ones. The Anaphe silk (common in Nigeria), the Kalahari silk (common in Namibia), and Borocera Madagascariensis (common in Madagascar) are examples of wild silks of Africa (Slegtenhorst & Venter, 2009).

Degumming

The cocoon is a matrix of silk fiber reinforcement within a gummy matrix. The gummy matrix is the sericin that surrounds the silk fibroin (Costa et al., 2018). While this structure serves a protective function to the silkworms, humans require the silk to be separated from the gum to be useful for textile and other applications. The degumming process essentially dissolves away the gummy matrix while leaving behind the silk fiber. The moths are allowed to complete their transformation and leave the cocoons. The cocoons are then soaked in water to separate and wash off residues. In one method, wood ash is added to water in which the cocoons are emersed and heated above boiling temperature. The extraction time can vary from 8 h or longer. An alternative method is to boil in 3% sodium carbonate and then in 3% soap solution. More recent methods where silk fibroin is being extracted for more advanced biomedical applications, degumming the cocoons by boiling in 1 M sodium carbonate prior to dissolving the fibroin in 9.3 M lithium bromide followed by dialysis to get purified silk fibroin (Costa et al., 2018). This is then followed by rinsing several times until the sum residues are completely removed.

Yarn Spinning

The mass of silk fiber obtained is then carded to detangle the fibers. This is then followed by spinning the fibers into yarn. The spinning process involves twisting multiple fibers into long yarns. The spun fibers are stronger, thicker, and more processible than individual fibers. The spinning is done manually by wetting the hands with water. The yarns are rolled up on spindles or rods ready to be sold to weavers. The yarns can then be dyed or used in their natural color.

Weaving

The weaving process follows the same fundamental technique as any other weaving process. In the pre-industrial age of Africa, this is mainly done manually using wooden weaving tools and setup. The yarns are laid out into webs that are arranged based on the desired patterns. In Nigeria, where the production of textile from silkworms was once predominant the spun yarn is woven into a highly valued fabric called aso oke. The weaving process involves weaving the spun silk yarns into long strips that are approximately 4 inches wide and 10 cm wide and 12 m long. These are then rolled into bundles and sold to tailors who then cut and sew them into desired garments.

To reduce cost, it has been common practice to mix cotton yarns with silk yarns in the making of garments in West Africa, particularly in the production of the aso oke in the Yoruba land of Nigeria. Today cotton and other synthetic textiles are mostly used in the production of aso oke in Nigeria. The anaphe silkworm silk called Sanyan is mostly reserved for ceremonial purposes as the raw material becomes rarer and less in demand. These methods of silk production are still being used in parts of Africa. Although the use of silk has been largely replaced by cheaper fibers like cotton and synthetic threads like nylon, the process for making the fabrics still remains. As the world looks for more sustainable processes and alternatives to synthetic fabrics, perhaps it is worth revisiting these processes and naturally occurring polymers.

One other interesting practice in the early textile industry of Africa was to unravel imported textiles and make use of them in the local textile styles which customers preferred. For example, in the Benin and Yoruba region of Nigeria, imported red woolens were unraveled and the yarns were used for brocades. Similarly in the Gold Coast (Ghana) imported silk textiles were unraveled and used in the "kente" fabric.

Production and Processing of Sisal

Sisal are described as one of the strongest natural fibers in the world (Torres et al., 2017). The main sisal producers in the world are Brazil and three African countries, Tanzania, Kenya, and Madagascar. It grows in lower quantities in other parts of the world like South Africa, Mexico, Haiti, Mozambique, and China (FAO, 2017). Sisal gives the longest spinnable fiber and highest density when compared to other fibers like kenaf, flax, hemp, ramie, and jute. For example, sisal has the longest spinnable fiber length of 500 to 1000 mm and a linear density of 12–20 tex while that of jute is 60–150 mm and 1.5–4.5 tex, respectively (Yu, 2015). The application of sisal includes mattresses, twines, ropes, cushioning for furniture, sacks, bags, floorings, construction, and pulp production.

The use of sisal in many of these applications has been replaced by synthetic fibers and other materials over the years. For example, use of polyurethane foams in matrasses and furniture cushions. Sisal fiber is still desired in applications such as fashion bags. As of 2015, the production of sisal by the four main producers was estimated at 16,000 tonnes. Across Africa, the production varies from handmade fibers cultivated on small-scale subsistence farms as found in Mozambique, to intensively farmed sisal grown on thousands of hectares of farmland as found in Kenya, Madagascar, and Tanzania (FAO, 2017). Other fibers are Kenaf, flax ramie, jute, hemp, pineapple (Yu, 2015), coconut coir, tinsel, cotton, sugarcane bagasse, bamboo, and banana.

Early industries in Africa largely involved in the harvesting of the raw fiber from producing countries and then shipping to more developed economies such as North America and Europe where they are then processed into products. Often these products are then exported to other markets (Shamte, 2000). The methods of production of sisal fiber have remained unchanged in many parts of Africa for over 5 decades.

Types of Sisal

Sisal have been classified according to different criteria. One classification is based on their sources. The lake/hedge sisal grows in East Africa around Lake Victoria. This is also classified as the estate-grown sisal. These are grown on large acreage intended for large-scale fiber production.

Sisal also varies by the production method. There is the manually decorticated sisal fiber and the mechanical water decorticated sisal fiber. The latter results in a product of higher fiber quality. Sisal produced in the main sisal-producing country of Africa is primarily produced using the mechanical water-based decortication method.

Sisal is also characterized by the length of the fiber: long fiber, medium fiber, and short fibers (Phologolo et al., 2012). The short lower grade fibers are used in pulp and paper, the medium and long fibers find use in applications such as rope and carpet production.

Cultivation and Harvesting of Sisal Plant

The sisal plant, Agave sisalana, is a hardy drought-tolerant plant that requires low maintenance. The plant has the ability to capture atmospheric moisture and its extensive root formation helps prevent erosion. It grows all year round in Africa and has hardly any known disease or pest. Sisal plants grow to around 15 cm within the first 9 months of their life. Within 2 years it attains a height of around 0.6 m. The leaves are the main part of the plant. They are long with the spiral arrangements. They are on average between 60 and 120 cm in length, 10.2 and 20.3 cm in width, and 2.5 and 10.2 cm in thickness. The leaves develop from the trunk in the center (Department of Agriculture, forestry and fisheries SA, 1015). The long fibers make up around 2% of the entire plant, the rest is discarded as biomass waste. The waste from sisal has potential applications in animal feed, energy, soil improvement, pharmaceutics, polymer composite, and other applications (Chand & Fahim, 2021). The plant itself when grown as hedges is a biological barrier against animals and intruders onto farmlands thanks to its strong elongated structure (FAO, 2021).

Cultivation of sisal requires no use of fertilizer or pesticides as it is a very resistant plant. It can be found growing wild in field hedges or along railway tracks (Chand & Fahim, 2021). It can grow in all soil types and even in saline conditions. Much of the sisal grown in major producing countries like Kenya gets exported as raw fiber. In Kenya, this is around 80%. They are used as raw materials for further processing into other end products. China is one of the major importers of sisal fiber.

Harvesting is done by hand. The leaves are cut off from the parent plant leaving sufficient leaf area for the plant to regrow new leaves over its lifetime. Some leaves are left uncut while harvesting. Depending on the age of 15 to 35, leaves are left uncut during harvesting. The number of leaves removed per harvest increases as the plant gets older (Department of Agriculture, Forestry and Fisheries, South Africa, 2015). The sisal plant is ready for harvesting the leaves after 24 months. Leaves can be continuously harvested for another 12 years of the plant's life. On average a sisal plant will yield 200 leaves during its lifetime. The growth conditions have an effect on the properties of the sisal obtained from the plant. Since the growth conditions determine how much lignin, hemicellulose, and cellulose are produced and their structure, the mechanical properties of each plant also depend on the age of the plant at the time of harvest (Phologolo et al., 2012). Figure 3.2 summarizes the process for sisal fiber production.

Fig. 3.2 Flow chart of Sisal fiber production process

Decortication of Sisal

This is the process whereby the plant is segregated into fibers and residue. The firm rigid flesh is separated from the moist pulpy part of the plant. The bast fiber of the plant is further separated into individual long fibers, short fibers, and residues. This can be done by mechanical separation or hand stripping. In another method, the fiber is separated using the impact stress of a machine without retting (Munder et al., 2008). The decortication process separates the useful fiber from the rest of the plant leaf. About 5% of the sisal leaf is long fiber (Lima et al., 2014), however only about 2% is obtained after processing. Decortication should be carried out within 48 h of harvesting.

Retting and Beating

The decorticated fibers are then retted. This involves immersing in water for about 7 days. This process weakens the components which hold the fibers together. This is followed by beating the leaves on stones, a process that removes residues and further separates the fibers. The residues are most often discarded as waste. Researchers have explored different ways to convert this residue from fiber production into useful products such as biomass energy, fertilizers, and composite reinforcements.

Drying

Following the beating stage, the fibers are washed to further remove residues. The fibers are hung to dry. Bleaching occurs during the sun-drying process when the fiber assumes its natural light brown color. Excessive sun exposure can cause color deterioration therefore this should be moderated. At this stage, the long fibers are selected by visual inspection and separated from the rest of the mass (Savastano et al., 2009). The fiber should be dried to at most 12% moisture content. Higher moisture content leaves the fiber prone to stiffening and matting as well as accelerated degradation (Department of Agriculture, Forestry and Fisheries, 2015). Preference is placed on long fibers with doing damage. These can be more effectively woven into products and fetch the higher market value. The shorter fibers can be used in other applications such as Plaster of Paris. To further detangle and smoothen the fibers, they are then combed and brushed. This can be done manually or mechanically. The fibers are also graded to separate them into specific ranges on basis of length, color, and other characteristics.

The processed sisal fiber can then be processed into various forms. Locally they are spun into yarns and then woven into products such as mats and sacks. In the Kenyan industry from 1914 to 1960s, a spinning factory was established in Juja during the period where sisal production saw the highest growth. Sisal can be blended with other fibers such as wool and acrylic either to achieve the desired features or to manage cost. Sisal fiber production consumes around 100 cubic meters of water per tonne of fiber produced. Thus, it is regarded as a water-intensive process.

Biodegradation of Sisal

Sisal, like many other natural fibers, is biodegradable and photodegradable. When exposed to high temperatures and/or sunlight, this tends to increase the degradation rate. When considering the environmental implications, this can be said to be a desirable attribute. In terms of application, biodegradation can prove restrictive. Sisal tends to be hygroscopic, it can draw moisture from the environment. The survival of the plant in drought conditions can be attributed to this moisture-absorbing property, however in the application of the sisal fiber this might be undesirable. As the fiber absorbs moisture, this encourages microbial degradation. The mechanical properties are also affected by increased moisture content which can result in deformation of the product as a result of fiber shrinkage or expansion (Chand & Fahim, 2021).

Production and Processing of Cotton

Textile production is reported as one of the most important manufacturing endeavors of the early industry of Africa. The other is iron smelting and smithing. The Sokoto Caliphate, now north-west and north-central Nigeria, was regarded as the biggest center for textile production and trade (Austin et al., 1871). Cotton production was an essential part of the economy in precolonial Africa. The value chain extended from farmers to spinners to weavers, dyers, tailors and embroidery makers to traders. Cotton once dominated the textile trade, especially in the trading ports along the coasts. The production processes and quality of cotton prior to contact with Europeans were determined by the taste and preference of the African consumers. These were influenced by factors such as culture, climate, function, and traditions. For example, in the areas around the Senegal River in the mid-1400 s, cotton garments were said to have been worn mainly by chiefs and other prominent members of societies while commoners wore mainly garments made of goatskin. Early records suggest that a thriving woven cotton market already existed in 1526 in Timbuktu (Kriger, 2005).

Cultivation of Cotton

Several cotton varieties exist and they vary in their properties. For example, in 2015, in Egypt, ten varieties of the cotton crop were produced and seven of these were used in commercial production (Hashima & Elhawary, 2021). Cotton picking involves separating the opened white fluffy cotton balls from the plant. This is done by hand or by machine (Negm & Sanad, 2020). Cotton was intercropped with other plants like maize, sorghum, millet, peanuts, yam, and rice. Cotton was sewn in typically early in the rainy season between June and November and within the first year the cotton plant, a shrub, would yield some cotton. The yield peaks in the second and third year and the plant continues to yield cotton for another 4 to 5 years after which the yield begins to decline (Bassett, 2006). The soil is either left to regenerate or other plants like legumes are planted next.

The growth stages of cotton are emergence of the cotton plant from seed, seedling development, squaring of the plant, flowering stage, setting of the cotton boll, and opening of the boll. The boll opening stage is the main stage where the desired product is formed on the plant. The cotton boll starts off as a closed green boll, this then gradually turns brown and eventually opens up revealing the white cotton boll made up of cotton fiber (Fahmy and Mobarak, 1971). This can take between 180 and 200 days for the cotton ball to open up completely ((Zhang & Dong, 2020). The conditions of growth such as soil type, nutrients, and sunlight affect the cotton yield and quality parameters such as strength, fineness, and length of the cotton fiber.

Records suggest cotton had been grown in Africa long before contact with Europeans. This is evidenced by the existence of so-called old world cotton species such as G.arboreum and herbaceum in places such as Nigeria, Lake Chad, Senegal, and regions of the upper Niger. Much of the cotton plants and products present in west Africa today are the so-called new world cotton plants that were introduced voluntarily during the Atlantic trade era and more were introduced by imposition during the colonial era (Kriger, 2005). Records suggest the earliest specimen of cotton fiber in Africa was found in Nubia, the tropical Nile Valley around 2600–2400 BCE. However, the evidence suggests that the cotton plant might have been used for animal nutrition rather than for textile production. Later evidence in the form of threads and woven fabrics of cotton found on mummies in the second-century BCE suggest that the cultivation and processing of cotton dates back three millennia from third-century BCE to fourth-century CE in Africa.

Other textile materials existed like textiles made from camel hair and non-woven textiles from flattened bark that were made by shaking and pounding of some specific tree barks (Kriger, 2005). Nonetheless cotton was the most widely made and traded.

Ginning

The picked cotton still has the rest of the seed stuck on them. This needs to be separated. Today, this process is done mechanically using machinery called the cotton gin (Negm & Sanad, 2020). In the early industries, they are thought to have been done by hand and preference was given to cotton plant species that could be ginned with ease. The process involved simply pulling out the white fluffy cotton from its base that serves as an attachment to the plant. Later simple tools were introduced for ginning.

Carding, Cleaning, and Combing

Cotton is produced as a tangled fluffy ball (boll) made up of cotton fiber. In this form, the cotton will have some residues and dirt on it. The cotton boll is cleaned to remove the dirts and residue, it is then opened up and straightened. The natural folded boll form needs to be arranged in parallel to each other as much as possible to facilitate further processing. The carding process makes use of a tool called a carder. The hand carder as known today was introduced by the Colonial Cotton Association alongside manual cotton gin tools at the turn of the century to some rural African cotton farmers and processors (Bassett, 2006). The carding and ginning tools achieved the same purpose with increased productivity. Today, the whole process from carding, and cleaning to combing is largely done by machines although small-scale handicraft textile processing is still done.

Spinning

This stage involves twisting the cotton fibers into yarns. Properties of the spun yarn vary depending on factors such as the spinning tension and the method of treatment and spinning. For example, Egyptian yarn toughness varies from 6.6 MPa to 5.41 MPa (Hashima & Elhawary, 2021). The quality of the spun yarn is a key determinant of the properties of the final fabric. The spinning of cotton was done manually in West Africa, this was aided by a spindle that is typically made of a wood shaft or ceramic spindle whorl. This is attached to a heavy object to provide balance for the spinning process. Spindle whorls were made out of baked ceramics from around 700–1200 CE around the Lake Chad region. Similar cotton spinning evidence has also been found around the Niger delta area between 900 and 1600 CE and the Middle Senegal Valley between 1000 and 1100 CE. There were handcrafters who were dedicated to the spun yarn trade. The spun yarns would be sold on to the weavers who then wove the yarns into fabrics.

Studies reveal that the spinning of fabric was done in the s-direction until the tenth century. However, after this period, spinning was done in the z-direction (Kriger, 2005). While the reason for this transition is not known, it can be said that the spinning technique was practiced very early on in African history from around 2000BCE and that the weavers sought to improve or change their methods.

Cotton Weaving

Weaving is the process whereby the yarns are formed into wefts and threaded warps. These are crossed over each other on a loom. The loom is a piece of machinery that aids this weaving process. Historic records indicated that vertical looms and treadle looms in different parts of Africa in its early industry (Kriger, 2005).

Evidence of cotton weaving date back to the eleventh and eighteenth centuries in Africa. Fragments of treadle loom used for cloth weaving were found in West Africa. Woven cotton was even said to have been used as some sort of currency around the lower Senegal River Valley Area (Kriger, 2005). In Angola during the period between the late 1880s and early 1900, yards of clothes were used as currency in exchange for rubber and other goods and services (Ball, 2000). These clothes most likely included cotton.

Fragments of fabrics found at an excavation site in Benin lead to the suggestion that in the mid-thirteenth-century fabrics made from a mix of cotton interwoven with other flat grass-like fiber were being practiced and the weaving was possibly done on vertical looms using spun fibers combined with non-spun fibers. Evidence of a variety of techniques used in textile products such as open work insertions, textiles with ribbed effect and structural patterning in woven fabrics, the combination of spun and unspun fibers, and varying weave density.

Records from travelers to the Gambia River region reported the cotton textiles that were made in this region were of impressive quality and quantity. They were mostly plain white, loom patterned, or patterned with stripes of blue- or red-dyed yarns. Similar reports of cotton garments being worn in Sierra Leone and the Rio Grande around the 1490 s.

The vertical looms were said to be common in the cotton production and export market in the Bight of Benin in West Africa. These types of looms were used to create long strips of fabrics which are then sewn together to make wrappers and mantles (Krigen, 2005). These were often dyed indigo blue if not plain white or loom patterned with blue and red stripes of dyed thread. In the upper Guinea Coast and then Cape Verde, the weaving of cotton was done on treadle looms.

Cotton textiles of various styles were produced in the early industry of Africa, they went by different names depending on factors such as the type of dyes used, weaving style, and where they were made. Examples are "Allada", "Benin cloth", "Guinea cloth", "Xereos", "Barafas", and "galans" among many other names for different cotton textiles. The technology to convert cotton plant harvest to woven fabrics in West Africa is thought to be over a millennium old in Africa (Kriger, 2005). Its origins in Africa have been traced to the upper Niger, Senegambia, and the Lake Chad areas. From there it spread to other parts of West Africa.

Production and Processing of Rubber

Rubber makes for an interesting plant as rubber trees have played an important role in African societies. For example, in Benin, a state in Nigeria in Western Africa, the introduction of the Para rubber species between 1879 and 1960 had a significant impact on land tenure, acquisition, and rights (Fenske, 2014). The traditional methods of allocation of land still remain relevant in many parts of Africa today. From the 1890s and onwards, rubber was one of the main products that was exported from East Africa. In many parts, the trade in rubber built on the pre-existing trade routes and networks that were already established from trade of other products such as palm oil, rice, and ivory (Monson, 1993). Export of rubber from producing countries to rubber-consuming countries rose from an estimated 3000 tonnes to 45 000 tonnes between 1860 and 1900. Prior to this period, the demand for rubber was negligible. Possibly because there weren't that many uses for rubber until 1840 when the process of vulcanization of rubber was discovered by Charles Goodyear who found that heat rubber in the presence of lead and sulfur resulted in more heat-stable rubber. This discovery fueled the growth in the rubber industry as vulcanized rubber found a much wider range of application. During this period rubber became so important in some economies that there are reports of rubber acting as some sort of currency where people received payments for labor in form of rubber credits (Monson, 1993).

Cultivation of Rubber

Rubber grew wild in the interior of many parts of Africa such as Tanzania, Nigeria, Congo, Angola, Ivory Coast, Cameroon, Ghana, Mali, liberia, Central African Empire, Senegal, and Gambia (FAO, 1977). It grows either as a giant vine growing around other trees or as a tree. It has dark green leaves and yields white sweet scented flowers. The fruit is round and yellow made up of a round hard shell encasing the seeds within a soft pulp reddish in color (Ball, 2000). Rubber tapping took place in the period of the year when rainfall was minimal in east Africa, (Monson, 1993). From 1910 onwards in parts of East Africa, rubber cultivated on plantation overtook rubber sourced from the wild.

Rubber species and quality varied from region to region. In East Africa in the inner Kilombero Valley in Tanzania, the Landolphia species grew wild in the forests at the hill foot. The rubber species grew as vines which could vary in diameter between a couple of centimeters to around 15cms. The trees took around a decade or a decade and a half to reach maturity before they can be tapped. Species in Brazil matured and were ready for tapping within 3 to 5 years.

Initially, much of the rubber was cultivated in Brazil where the Hevea brasiliensis was found. As global demand rose, particularly in the wake of the invention of the pneumatic tyre in 1880, other sources of rubber were sought to meet the demand. Hence, there was a mass exportation of rubber in countries such as Angola, Congo,

Mozambique, and Madagascar began. The main species found in these regions were the Landolphia and Clitandra. Other species included Ficus vogelii and Funtumia elastica. Later rubber began to be sourced from West African countries like Sierra Leone, Gold Coat (now Ghana), and then Lagos in Nigeria. In Lagos, Agege Area in mainland Lagos and Ota on the outskirts were once locations of agricultural plantations where some rubber was grown. Much of these are now urban areas.

Some records show that some technology transfer was facilitated by the colonial administrators, for example, between Fanti rubber tappers from the Gold Coast and the tappers in Ibadan, Nigeria (Omosini, 1979).

The Rubber Tapping Process

One of the earliest known applications of rubber was described in 1530. Rubber was said to have been used in an Aztec game where it was used as a bouncy ball. The latex was collected from the tree and rolled up into a ball. The Indians of Haiti were also recorded to have used rubber in this way. There are also records of how Indians used rubber in making crude footwear and bottles by coating the rubber sap unto their cloaks and earthen molds, respectively, and allowing them to dry (Ball, 2000). Therefore, some crude processing of rubber can be said to have been done before the 1500 s.

The process of extracting rubber from the rubber tree is referred to as tapping. There are a variety of methods for tapping rubber. The techniques also vary from tree to tree partly because the manner in which trees produce sap varies. For example, the sap of the landolphia species tends to dry up quicker. A method widely practised by the Fantis of the then Gold Coast and adopted by the Ibadan tappers in Nigeria (Omosini, 1979) is known as the Herring bone method. In this method, a long cut is made on the tree bark serving as the main cut. This is then followed by making subsidiary lateral long narrow cuts in slanted formation. The rubber tree sap referred to as latex which emerges as a result of these grooves is then collected in containers. These are then prepared into rubber lumps which are also referred to as rubber biscuits.

In one method of tapping rubber that was said to have been used for tapping landolphia rubber near the Congo River, (Ball, 2000) this method involves making cuts of bark slices from the trunk and branches of the rubber tree that are between 3 and 10 inches long and 0.5 and 0.75 inches wide (Omosini, 1979). The milk/sap that is then exuded from the cut is immediately collected with the fingers and spread unto the forearm and/or other body part to dry. Once dry, the rubber is peeled off. This is then boiled in water for the rubber to coagulate. In another method, vertical cuts are mad in the form of strips. Salt water was then splashed unto the cut. This causes coagulation of the rubber as it is exuded from the tree, it was then collected by rolling it up into a ball. Other methods involved making circular cuts of 2 inches in diameter on the bark. Several cuts were made at 18 inches intervals, the tappers then used salt water and lemon juice to coagulate latex. This method is called the ring method as it results in the formation of rings of coagulated rubber if the latex is left for several hours to form hardened coagulated rubber rings on the vine. The rings are then rolled up into balls.

Towards the end of the rubber boom period in Angola, around the late 1880s to early 1900s, the rubber trees were being depleted and the tappers adopted the more crude methods. In one method, the stems of the rubber plant were macerated in water to allow the rubber to coagulate within the bark (Ball, 2000). This is then dried in the sun. The stem is then cut into shorter length followed by beating which loosens the bark. The bark is then stripped off followed by further beating with sticks and stones to remove the wood debris leaving behind the lump of rubber which had coagulated in situ. More boiling and beating follows to further remove debris. Due to the crudeness of the method the rubber obtained is of low quality with higher level of impurities compared to other methods.

Other techniques for tapping rubber from rubber trees include Spiral, v-shape, and vertical methods. The method chosen depends on the skill of the tapper and the type of rubber tree. Most of the methods used in the early rubber industry of Africa were destructive to the tree. Some methods required the stem to be cut to remove the bark which was then pounded to extract the rubber. For every rubber tapping method, the tree should be given a period of recovery. This varies from 6 months to several years (Omosini, 1979).

Processing of Latex into Rubber Lumps

After the latex is collected from the tree, these then need to be processed into forms that are easier to store and transport. These are referred to as lumps or biscuits. In Lagos, Nigeria the cold methods or the heat methods were predominant at the time. The former originated from the Fanti rubber tappers while the latter originated from the Yoruba rubber tappers. The cold technique involved placing the latex in the trunks of fallen trees. These were then covered with leaves of the palm tree and stored for several days in this form. The water is gradually evacuated from the latex by evaporation and absorption into the trunks. This leaves behind a coagulator rubber mass. To remove the remaining water, the rubber mass is rolled with a wooden roller or a bottle serving as a roller. The rubber lump formed using this technique is typically a lighter brown on the outside with a darker brown center.

The heat method required evaporating the water from the latex heating in boiling water. This resulted in the coagulation of the rubber. The water is then removed by straining. The coagulated rubber is then dried. This method results in a rubber product with color closer to black with a sticky texture. The rubber from the heat method was regarded as inferior.

In a third method, rubber biscuits are produced through a chemical process that results in flattened dry "biscuits" of rubber. Chemicals used included acetic acid, potassium carbonate, formalin, and alcohols. This method was much less used in the early industry of Africa.

Transporting the Rubber to the Trade Point

The rubber produced then needs to be transported to the ports where they are bought by European traders who then ship them abroad to be used in the manufacturing of rubber goods. During this period the railway was unavailable in many parts. In Ibadan in Nigeria, for example, the rubber, like many other goods produced from these areas for export, is then transported by porters who would load the goods on their heads from Ibadan to Lagos (Omosini, 1979). Today a train service does exist between Lagos and Ibadan as of the year 2021. The use of animate energy to power transportation and processes was more prominent during this period. Push carts powered by people or horses are still seen in some cities across Africa. For example, Fig. 3.3 is an example of a horse used for potted plant delivery in Dakar, Senegal.

Environmental Impact of Rubber Tapping in Early Industry of Africa

There are reports of high levels of deforestation due to unsustainable rubber tapping methods. For example, Morrison (1993) reported the extent of violation of regional forest ecosystem in East Africa caused by the rubber trade. During the period of the 1890 and early 1900s, rubber trade thrived in East Africa. Similar reports of degradation of the forest in Ibadan and Lagos were mentioned in the paper by Omosini (1979). In Ibadan this led to banning of rubber tapping by the authorities for a period which ultimately resulted in a switch of interest from rubber trade to cocoa trade.

Rubber tapping can be done in a less destructive manner that does not require cutting of the rubber tree. However, regardless of method used for tapping, if not period is given for the tree to recover from the tapping, the process is unsustainable. Lack of regulations and desperate attempts to obtain large quantities of rubber quickly without consideration for the environment motivated unsustainable rubber tapping methods. The impact of the rubber trade is still felt well into the 1990s (Monson, 1993).

Ivory Sourcing and Processing

Ivory refers to the material the elephant tusk is made of. It is also known as dentin and is essentially a giant tooth with an outer layer. It is a biomaterial that is made up of mineralized collagen fiber in specific arrangements known as Schreger pattern (Alberic et al., 2018). The tusk is long tapered structure. The tooth pulp which comprises living tissue is present in the hollow of the tusk. The tusk has similar physical properties as a tooth but with more flexibility and strength. Elephant makes

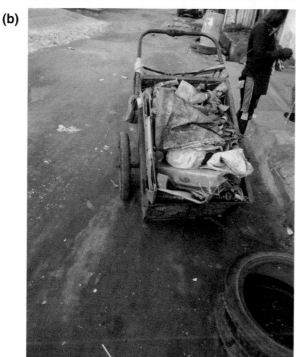

Fig. 3.3 a Horse cart used for delivery of potted plants in the city of Dakar, Senegal, December 2019 **b** a push cart the owner uses for collection of resellable waste. Lagos, Nigeria June 2022

use of the tusk for various purposes such as digging, lifting, and defense. Elephant tusks are desired for their fineness, luster, strength, and relative ease of processing. Ivory has been associated with names like "white gold" and "plastic of the nineteenth century". This is due to the plastic property the polymer demonstrates and the fact that much of the products, such as billiard balls, piano keys, and combs, that were once made from ivory were later made using plastic as cheaper and elephant cruelty-free substitute.

Earlier applications of ivory included beads and ornament. They were also carved into figurines. There are several mentions of ivory trade in different parts of Africa in the early trade and industry. Evidence of ivory trade from the east coast of Africa date back well before the first millennium AD (Coutou et al., 2016). Ivory was exchanged for other goods such as cloth, beads, and copper. Ivory trade was at its peak towards the end of the nineteenth century. By the end of the nineteenth century Ivory trade was ended. In Malawi, for example, Ivory accounted for about 80% of total exports in the last decade of the nineteenth century. Although ivory trade has been banned since 1989 by the Convention on International Trade in Endangered Species agreement, the exemption of trade of some stockpiles and illegal poaching is still a problem.

The Ivory Trade

Tanzania, Mozambique, South Africa, Namibia, Zimbabwe, Botswana, Ivory Coast, Senegal, Ghana, Cameroon, the then slave coast (which included Nigeria, Republic of Togo, and Benin), and Guinea coast are some of the countries where Ivory trade occurred. There are records of the Atlantic trade of Ivory from 1600 (Feinberg & Johnson, 1982). In Malawi, for example, the ivory trade played a crucial role in the political history of the country up until the nineteenth century. Political powers were dependent on the control of the ivory trade in the mid-seventeenth century. Ivory was traded for firearms and these enabled groups to take over regions they fought for and for groups to rebel against centralized powers and hunt elephants. One record estimated firearms import into East Africa at 90,000 (Beachey, 1962). The ivory trade in East-Central Africa is said to date as far back as 1000AD. Archeological evidence suggests earlier use in Egypt where Tutankhamun is buried with a headrest made of elephant tusk. Much of these were shipped to India, China, and Europe. In the eighteenth century, the rise in the manufacture of goods, such as piano keys, ornaments, umbrella handles, chess sets, snuff boxes, and other goods which used ivory, resulted in an increased demand for ivory. Some publications suggest that the ivory shipped from North America and the West Indies and later shipped to Europe was sourced from Africa.

Between 1800 and 1989, the population of elephants declined from 26 million to 6 hundred thousand. The killing of elephants was once done for sport and the tusk was kept as a souvenir. As the market value for tusk rose due to its demand for manufactured products, the killings of elephants further rose.

Ivory from Elephants

Ivory of elephants grows throughout their lives such that the older the elephant is, the larger the tusk (Chaiklin, 2010). Three main species of elephants are known: the bush elephant, the savannah elephant of Loxodonta Africana, and the forest elephant or Loxodonta cyclotis both found in Africa. The third species of elephant known as the Asian elephant are found in South and Southeast Asia. All three species of elephants are all said to originate from Africa and like all animals with trunks and tusks, they are thought to be distant relatives of the Palaeomastodon which are said to have roamed the earth 40 million years ago. The African elephants are larger in size than the Asian elephants (Chaiklin, 2010). Other animals with tusks include walrus, hippopotamus, mammoths, wild boars, narwhal, and sperm whales.

Ivory is classified into soft and hard ivory. The ivory from the Asian elephant is the hardest and forest ivory is often the softest. However, the hardness of ivory of any given animal is dependent on the diet of the elephant (Chaiklin, 2010). The color of the tusks also varies widely. They can be pure white in color or all shades of yellow, pink, green, and brown.

Ivory Removal from Elephants

Although on some occasions surgical elephant removal is carried out for the health and safety of the animal. For example, in the case of a crack (Stener et al., 2003) or an infection (Welsch et al., 1989). These can be done surgically with great care and usually over a long period of 3 weeks to 1 year. Commercial ivory (now illegal) is taken from the elephant by force which causes the death of the animal. Elephants are large animals and therefore need to be killed to render them harmless. Also, forceful removal of the tusk means interfering with other crucial tissues which are fatal. In some parts, only the tip of the elephant tusk is removed. This can be done without any harm to the animal. Images of elephants with tipped tusks who are still alive and well are seen in images from the 1900s (CHaiklin, 2010).

Processing of Ivory into Products

Evidence of shavings and fragments found in KwaZulu-Natal of objects like bangles made of ivory that were between 18 and 27 mm suggested that during this period ivory was being carved as a processing method in the seventh- and eighth-century AD (Coutu et al., 2016). Although the trade in ivory was classified into worked and unworked ivory (Chaiklin, 2010), much of the ivory that was exported required little processing other than removal from the elephant and perhaps wiping the blood and dirt. Therefore much of the processing of ivory in the early trade was done by craftsmanship. The main tools used were rasps and chisels.

Ivory is a very tough and resilient material that can withstand a wide range of temperature, chemical, and mechanical treatments. Ivory does not decay, however over a period the color and other physical properties can change. Depending on the storage condition, the nonmineral components can dry out rendering the ivory more brittle and prone to cracking. Ivory is easier to maintain than wood. From the 1900s onwards, ivory was processed into different products mainly by machining using rotary drills (Chiaka, 2010). With the industrial revolution, products can be mass-produced using electricity and moving assembly lines that drove the second industrial revolution.

Production and Processing of Leather

The flaying of hides and skins by indigenous Africans is said to set the foundation for the leather industry. Animal skins were already worn as decorative clothing or as clothing reserved for noblemen or men of certain professions such as hunters and warriors. Certain animal skins such as those of lions and leopards were symbolic of power and prestige. Other than clothing hides and skins were used as carpets or wall hangings, amulets, charms, and other applications. Some of these are still used in certain African communities of present day albeit rare in modern Africa.

The animals from which hides and skins were obtained include cows, sheep, and goats, and more rarely reptiles and other wild animals. In Nigeria, the north was the center for animal hide and skin production. The trade of hides and skin preceded that of cotton in this northern region of Nigeria.

Raising of Animals for Leather

In the Hausa land in Nigeria, goats and sheep were raised by smallholder farmers and herders. The farmers also farmed crops such as groundnuts and cotton alongside raising animals. The Fulanis who are pastoral farmers raised larger flocks of sheep, goats as well as cattle. They grazed the land in search of water and pasture and lived a nomadic life. The main areas were Kano, Sokoto, Katsina, Zaria, and Bauchi emirates (Adebayo, 2020). The animals were sold at markets where they are bought for meat or rearing. The main use for the livestock was meat and leather was a by-product.

Extracting of Hides and Skins for Leather

The process involved slaughtering and then flaying of the animal to remove the hides from large animals and the skin from small animals. The flaying was carried out by many butchers. This involved removing the skin or hiding from the animal with a sharp knife and ideally with minimal damage to the skin. However, reports suggest that the flaying process done by the butchers in early 1900 often had stray knife marks which are undesirable. Later in the mid-1900s, the colonial authorities introduced some strict rules on flaying and drying in order to increase the export value of the hides and skins produced in this area (Adebayo, 1992). This required that where the skin is being removed from small animals, a knife is used only for the initial cut, and the rest of the removal is done by hand. In the case of hides, the flaying should be done with care not to cause any damage to the hide. The rest of the animal is used for meat and remnants for other applications like glues and fertilizers.

The next stage is tanning. This required specialized skills which were less common than flaying. For this stage, tannic acid was extracted from the Acacia arabica pods. The tanning process used in Kano is regarded as very high quality. After tanning the leather is dried. The drying was initially done by drying them laid out in the sun. Later with new rules imposed by colonial authorities in the 1930s, the drying is done by hanging the hides and skins on frames that allowed proper aeration and distribution of heat which minimized decay and uneven drying, and the leather is then dyed if required. The dying is done using extracts of the sorghum leaf sheath (Kayode et al., 2011). Later the market began to show a preference for undyed leather.

In Kano, the leather produced is referred to as "Morocco leather" and is regarded as the best in the world. The leather from these parts was used locally and also traded with Tripoli, Morocco, and then Europe. This trade route from Morocco to Europe is the reason why the leather from Kano is called Morocco leather. Kano is well known for quality leather goods such as sandals, bags, and shoes. Towards the end of the nineteenth century, the leather market of Kano began to decline. This is attributed to the switching of demand to the leathers made using the chrome tanning process and the increased colonial influence in Northern Nigeria.

Production and Processing of Gum Arabic

Gum Arabic has been harvested and traded in Africa for thousands of years. Gum Arabic is a hydrocolloid polysaccharide, characteristic chemistry is shared by all gums. Its backbone is a chain of 1,3-linked β-D-galactopyranosyl with other subsistent units attached. The protein content can vary between 1.5% and 3.0%. It also comprises minerals such as calcium, magnesium, sodium, potassium, iron, and phosphorus. A shared characteristic for all gum Arabic regardless of source is relatively high solubility in water at varying temperatures from cold to hot, and a relatively low solution viscosity compared to other polysaccharides. This gives it an advantage in applications where its properties like emulsifying properties can be harnessed without altering the viscosity of the formulation.

Cultivation of Gum Arabic Tree

Gum Arabic is also known as acacia gum. It is obtained from the acacia tree which grows in the wild across the Sahel belt of Africa from Senegal to Somalia. Countries, where the acacia tree grows, include Sudan, Somalia, Ethiopia, Niger, Chad, the Central African Republic, Nigeria, Senegal, and Mauritania (Awad et al., 2018). Sudan, Chad, and Nigeria are the main producers of gum Arabic. The tree grows to a height of around 17 m culminating in a flat crown-shaped top. In the eastern and western African region as well as the Arabian Peninsula, the acacia seyal variety is found (Awad et al., 2018). The acacia fistula is found in Eastern African regions. The seyal variety is recognized by its greenish-yellow to reddish-brown colored bark. The fistula varieties are identified by their greenish-yellow toned bark.

Extraction of Gum Arabic from the Acacia Tree

Gum Arabic is present within the trunk and branches of the tree. Harvesting is done on branches that are wider than 3 cm. To obtain gum Arabic, an incision is made on the trunk or branch. A portion of the bark is then torn out. This injury causes the tree to exude a watery sap that resembles latex. This exudate coagulates into large, crystalline balls with a waxy texture. In this solid state, they can then be picked from the trees. In some cases, Acacia seyal tree collections are made from natural exudates of the tree.

The properties of any given batch of gum Arabic vary depending on the season, tree species and location, and growing condition. The two main species of gum Arabic used commercially today are Acacia Senegal and Acacia Seyal trees. Generally, the Acacia Senegal is regarded as superior due to its colorless appearance and superior emulsifying properties (Olatunji, 2018). Typical annual yield from an acacia tree is reported as 300 to 700 g. Gum can be collected from a single tree up to three times a week giving a recovery period of 3 weeks in between collections.

Other Polymer Products

Polymers are ubiquitous, they can be found almost everywhere in nature and even many more have been synthesized. While it is not possible to cover every single polymeric material that was being processed and used in the early industries of

Africa, this chapter has attempted to touch on a few that have been of major economic significance. Here we touch on a few more worthy mentions.

Starch in Grains

Grains like millet contain between 51 and 79% starch (Annor, 2017). The other parts of the plants like the stalk contain cellulose, the most abundant polymers on earth. Take the case of Zimbabwe as an example. Millet and sorghum have been grown by the Shona people over an estimated period of 2000 years before the introduction of maize in the sixteenth century by the Portuguese (Tavuyanago, 2010). These grains are ground and prepared for different meals, nonalcoholic beverages, and alcoholic beverage production. Beer produced from the fermentation of sorghum is said to have a relatively high vitamin B content. The grains and the stalks of the crops were also used as animal feed. The stalks were used in the construction of buildings, barns, fencing, and for beddings. Although maize dominates much of the grain market in the colonial era and beyond, sorghum and millet plants grow faster, are more drought and pest resistant, and require less fertilization. Millet and sorghum grains are also easier to mill into powdered forms than maize.

Thaumatococcus Daniellii Leaves

Leaves contain cutin and suberin polymers alongside cellulose and other biochemicals. Certain leaves are valued for their bioactive properties and/or their physical properties. Thaumatococcus daniellii, for example, has been primarily used for cooking because of its physical properties and stability at high temperatures. It is commonly found in the rainforests in West Africa (Yeboah, 2003). It is used in the preparation of foods like beans cake where the beans that have been milled and formed into a liquid mix form with water and other ingredients added for flavor. The liquid beans mix is then poured into the leaves and placed in boiling water in a pot. The beans thicken into a bean cake as it is cooked in the leaf. The thickening is thought to be a result of the gelatinization of the polysaccharides in the beans. This method is still being used today as shown in the image of beans cake (moi moi) in Fig. 3.4 taken in 2021. The fruits of the plant are also used as sweeteners and/or taste modifiers (Yeboah et al., 2003). The leaves are also known to contain some bioactive compounds.

Polysaccharides in Gums of Terminalia Sericea Tree

Terminalis sericea is a tree that extrudes a honey-colored gum that tends to cling to the bark. The consistency ranges from a dry brittle solid mass to a runny liquid.

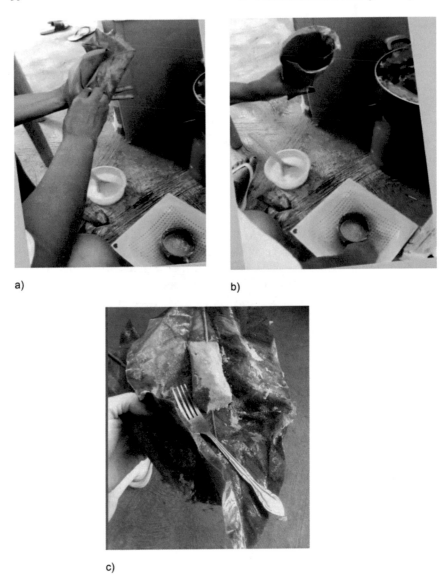

a)   b)

c)

Fig. 3.4 Making of *"Moi Moi"* (Bean cake) using Thaumatococcus daniellii leaves **a** the leaves are folded **b** the blended beans with spices are poured into the folded leaves and placed in the pot of boiling water **c** Prepared beans cake served in the leaves it was cooked in. Lagos, Nigeria November 2021

In the dry form, the powder is dispersible in water. It is known to have been used by a group of hunters in some parts of Namibia (Wadlet et al., 2015). The glue is used to hold together tools and weapons. Chemical analysis (Anderson & bell, 1974) shows that the polysaccharides present in the gum are said to be made up of sugars like glucuronic, galacturonic acid, 4-O-methyl glucuronic acid, rhamnose, arabinose, galactose, xylose, and mannose in complex arrangements. The gum is soluble in water and shows the viscosity property of polymers as it dissolves in water forming a thick viscous liquid.

Starch in Cassava

Cassava serves as a staple in many parts of the world. It is a root crop with a starch content of the root around 30% and between 80 and 90% carbohydrate content (Bantadjan, 2020). Take Uganda as an example. Cassava which originates from the southern border of the Amazon Basin (Olsen & Schaal, 1999) began to be grown in Uganda in approximately 1862 through the Tanzania-Arab trade and to date has become a staple food of high importance in the country. Cassava is processed in different ways depending on the end-use. In food production, it can be boiled whole, milled to powdered form for pastes and liquid meals and it can also be sliced and fired as cassava chips. Cassava is also used in the brewing of beer and distilled alcohol. In the early industry in Uganda, a company called The Lira Starch Factory (Otin-Nape et al., 2005) ventured into the production of cassava starch. It sourced raw cassava from farmers and extracted starch for industrial applications. The company operated between 1968 and the 1980s. Another root tuber yam is said to have partly originated in West Africa and has been domesticated since around 5000 BC (Andreas et al., BC). It is mainly used as a food crop that is still grown and eaten in modern-day Africa.

Baskets and Basketry Objects

Baskets have been woven in many parts of the African continent for centuries. Archaeological evidence and paintings on the walls of tombs show images of baskets getting used in ancient Egypt (Peek, 2004). The techniques used in basket making have existed for centuries and are still in use today. Being a versatile technique, basketry beyond the container basket finds a wide variety of uses. These include boxes, cases, trays hats, belts, pursed sacks, anklets, bracelets, doors, walls, fences, enclosures, granaries, chicken coops, fish traps, beehives, roofings, and other applications.

Baskets are made from a variety of materials which are mostly fibrous plant materials. Materials derived from animals like hairs, hides, and leather are also used in basket weaving albeit less often. In modern times, basket weaving using synthetic fibers has also emerged. Plant materials used in basket weaving include sisal, wood, grass, palms, bamboo, and reeds. These are natural fibers that are made up of natural polymers, cellulose, and lignin. Basket weaving requires fewer tools and no equipment unlike more complex processes like textile spinning and weaving.

Various basketry techniques exist such as the coil, plait, twine and stake, and strand. Certain techniques are common in different parts of Africa. For example, the sewed coil technique is more common In Ethiopia, Rwanda, Burundi, and other parts of Southern Africa. The twill method of basketry is more common in parts like Zambia, northern Mozambique, and countries along the East African coast. Techniques like open twine and stake and strand are the most common methods that are common in all parts of Africa. Basketry involves obtaining the fiber from the plant, which may involve different processes depending on the type of fiber used. For example, where sisal has used the process for obtaining the fiber is discussed in a different section of this chapter. The fiber may then be soaked in water in order to make handling and bending them during the process of weaving easier. Basket weaving is mainly done by hand with the aid of an awl and a knife as the basic tools used. Figure 3.5 shows examples of woven baskets in Abeokuta, Nigeria.

Calabash Making

Calabash is the dried hard often rounded multi-use item obtained from the fruit rind of the plant Lagenaria siceraria commonly known as the bottle gourd. It grows as an annual plant which can be a monoecious trailer or a climber (Konan et al., 2018). It is a diverse plant that varies across different cropping systems and environments. The calabash has been in use in several parts of Africa for centuries. It produces fruits that could either be edible or nonedible. The fruits of the Lagenaria Siceraria plant vary in shapes and sizes which could be a round, egg-shaped, cylindrical, elongated, dipper, or other shapes and contain seeds that are rich in proteins and lipids. The rind comprises lignin and cellulose (Konan et al., 2018). The nonedible variety forms a characteristic harder shell which makes it better suited for use as a calabash. The edible variety serves as a food source.

Calabashes are made into cups, containers, jugs, musical instruments, utensils, floats, fishing nets, decorated ornaments, and other items, They are mainly processed by carving, and post-processing may involve painting and adornment with beads or corals. The round-shaped calabash, for example, gets cut into two halves which serve as bowls and cups for eating or drinking. The texture is similar to that of carved wood. Although largely replaced by modern materials like plastics, calabash still gets used across Africa today for functional uses, aesthetics, and art. Figure 3.6 shows calabash growing on a calabash tree in the town of Idanre, Ondo State, Nigeria. Figure 3.7 shows the image of calabashes on display at a market in Abeokuta, Nigeria, and Fig. 3.8 shows calabashes being used to serve drinks as observed at an urban bar in Lagos.

Fig. 3.5 Baskets and basketry trays. Here one of the trays is used to place onions on display. Abeokuta, Nigeria, February 2022

Ancient Fibers in the Modern Era

Despite having been replaced by synthetic fibers in recent years, there is renewed interest in applications of natural fibers like sisal in high-value products such as reinforcement fiber in the floor and wall covers, reinforcements in building Plaster of Paris (POP), ornate ceilings, flower vases, brushes, and many others (FAO, 2017). Recent years have seen advancements in the production of sisal fibers for export with Saudi Arabia being an emerging major importer and Kenya a major exporter (FAO, 2017). China has been a major importer of sisal fiber for several years. Sisal is used in pulp and paper (sisal pulp has a desirably high cellulose content which makes it a good substitute for wood pulp). Its absorbent nature and higher resistance

Fig. 3.6 a Calabash fruits growing on a calabash tree in the town of Idanre, Ondo State, Nigeria. **b** Close-up image of calabash fruits growing on the tree. February 2021

Fig. 3.7 Various calabash types are on display at a market in Abeokuta, Nigeria, February 2022

Fig. 3.8 Calabash used to serve drinks at Afefeyeye bar and restaurant in Lagos, Nigeria. Such application is intended for nostalgic appeal although the palm wine itself is packaged and stored in plastic bottles. April 2022

to folding also make it useful in pulp and paper production. It also improves bulk in paper and cardboard (Department of Agriculture, Forestry and Fisheries, 2015), polishing cloth, pop, and sisal in injection-molded composite plastics/polymers. Its application as buffing or polishing cloth is due to its adequate texture that is moderately strong to polish surfaces like steel without scratching. Paper products from sisal have the desired porosity that makes them useful in applications such as tea bags and breathable wrapping paper. Sisal can substitute or augment the use of fiberglass reinforcement in plastics like nylon in applications like brake pads, roofing, and fiberboard. It is already used in dart boards.

In Tanzania, residue from sisal production is being used as biomass in the commercial generation of green electrical energy (Phologolo, 2012). It is also used in the commercial production of fertilizer and biogas. Studies show that 75% of the energy in the country generated from sisal by-products can be used in supplying homes and 25% is supplied to the sisal-processing industry (UNIDO and Tanzanian sisal industry fund on sisal). The juice from the sisal plant has been investigated to have some bioactive properties for pharmaceutical applications. Active components can be used in products such as inulin and hecogenin.

Sisal applications in ropes and twines have been widely replaced by synthetic polymers such as polypropylene and nylon. Within a time space of 14 years between the 70 s and 80, polypropylene replaced 55% of the sisal market share (Shamte, 2000). In Tanzania, sisal was once a major export commodity with 234 thousand tonnes produced annually in the 1960s. This figure declined to 30,000 tonnes by the middle of the 1980s (Hartemink and Wienk, 1995). This decline has been attributed to the emergence of synthetic fibers and the decline in soil fertility due to inefficient agricultural practices.

More contemporary application of sisal fiber includes use as a composite material. In this application, it serves as the fiber within a polymer matrix of synthetic polymers such as epoxy, polyethylene, polyester, and phenolic polymers (Chand & Fahim, 2021). Sisal is one of the fibers that is being explored to meet the demand for more sustainable and greener buildings. One of such ventures is the production of cement-based composite using sisal residue (Lima et al., 2014).

As African countries such as Rwanda, Kenya, Ethiopia, and other countries across the world move towards implementation of bans and other plastic reduction policies, increased demand for natural fibers such as sisal is anticipated. These can substitute plastic products such as ropes, baskets, and woven bags.

The price of sisal reported an increase from 750 USD a tonne to 1,010 USD a tonne between 2003 and 2006 in Africa. In Brazil, the price increased by 380 USD from 400 USD in 2002 to 780 USD by the middle of 2006 (Savastanojr et al., 2009). 2003 saw the beginning of an improvement in the global demand for sisal and African sisal has always fetched a higher price due to its superior quality although Brazil is the highest producer, producing 139,700 based on 2014 figures. This can be attributed to increased demand for sustainable products and a shift from plastic-based fibers. With rising concerns from issues such as microplastics from textiles made up of plastics like polyester and the depletion of fossil reserves, renewable and biodegradable options like sisal and other natural fibers have become more attractive as they have proven to be renewable and reliable over centuries of production.

Cotton processing has advanced from hand-spun cotton woven on vertical wooden looms to finer cotton fabrics spun and woven by machine alongside other processing stages such as roving and drawing. Recent studies look into the modification of fiber strength through genetic engineering (Zang et al., 2021). Some of the issues associated with current methods of high-volume processing of cotton textiles include the health risk factor for workers. Studies on cotton industry workers in Egypt indicate that these cotton workers face the risk of developing respiratory health complications and this is prevalent in those employed in spinning and weaving (Tageldin et al., 2017). Today, one of the modern applications of cotton is in the production of denim. The weaving of denim employs innovations such as intricate weaving and yarn blending technology to achieve the trademark properties of denim (Rania et al., 2015).

3D printable synthetic alternatives to ivory can potentially replace lost artifacts made of real ivory (Rath et al., 2021). New studies have presented UV-polymerized calcium phosphate-filled dimethacrylic resin that produces properties resembling those of ivory from elephants. However, there are synthetic alternatives that can serve as a more feasible option in terms of production quantities and available resources. Silk fibroin is now being explored for biomedical applications such as the production of microneedles for drug and vaccine delivery (Lu et al., 2022), bio-inks for 3D printing of tissue scaffolds, and recently they are being studied for potential application in flexible electronics and sensing devices (Moreira et al., 2022).

Figure 3.9 shows an example of natural fibers still used for building construction in Dakar, Senegal. This illustrates the old uses of these natural fibers, although largely replaced by plastics, they are still used for aesthetics and functionality in the modern day.

Conclusion

Diverse techniques were employed in processing different polymer-based materials in early trade and industry of Africa. The raw materials processed were sourced from both plants and animals and largely made use of manual labor and skill. Some simple machinery and tools such as carders and looms were introduced in this early industry. The processing added value to the raw materials and served as a means of trade.

Fig. 3.9 Natural fiber used for ceilings and windows of a restaurant in Dakar, Senegal, December 2019

References

Adebayo, A. G. (1992). The production and export of hides and skins in colonial Northern Nigeria, 1900–1945. *The Journal of African History, 33*(2), 273–300.

Alberic, M., Gourrier, A., Wagermaier, W., Fratzl, P., & Reiche, I. (2018). The three-dimensional arrangement of the mineralized collagen fibers in elephant ivory and its relation to mechanical and optical properties. *Acta Biomaterialia, 72*, 342–351.

Anderson, D. M. W., & Bell, P. C. (1974). The composition and properties of the gum exudates from Terminalia sericea and T. Superba. *Phytochemistry, 13*(9), 1871–1874.

Austin, G., Frankema, E., & Jerven, M. (1871). Patterns of Manufacturing growth in sub-Saharan Africa. The Spread of Modern Industry To The Periphery Since 1871. Oxford University Press, pp 345–381

Awad, S. S., Rabah, A. A., Ali, H. I., & Mahmoud, T. E. (2018). Acacia Seyal gums in Sudan: ecology and economic contribution. In: Gum Arabic: Structure, Properties, Application, and Economics. Marios, A. A. (Ed.). Elselvier Accademic Press. UK, pp. 3–5. ISBN: 978-0-12-812002-6.

Ball, J. (2000). A time of clothes: The Angolan rubber boom. 1886–1902. *Ufahamu: A Journal of African Studies, 28*(1), 25–42.

Bantadjan, Y., Rittiron, R., & Malithong, K, Narongwongwattana. (2020). Rapid starch evaluation in fresh cassava root using a developed portable visible and near-infrared spectrometer. *ACS Omega, 5*(19), 11210–11216.

Bassett, T. J. (2006). *The peasant cotton revolution in West Africa: Cote D'Ivoire, 1880–1995* (p. 33). Cambridge University Press.

Beachey, R. W. (1962). The arms trade in East Africa in the late nineteenth century. *The Journal of African History, 3*(3), 451–467.

Chaiklin, M. (2010). Ivory in world history- early modern trade in context. *History Compass, 8*(6), 530–542.

Chand, N., & Fahim, M. (2021). Sisal-reinforced polymer composites. Tribology of Natural Fiber Polymer Composites (2nd ed). Woodhead Publishing Series in Composite Science and Engineering, pp. 87–110.

Chen, S., Liu, M., Huang, H., Cheng, L., & Zhao, H. P. (2019). Mechanical properties of bombyx Mori silkworm silk fibre and its corresponding silk fibroin filament: A comparative study. *Materials & Design, 181*, 108077.

Costa, F., Silva, R., & Boccaccini, A. R. (2018). Fibrous protein-based biomaterials (silk, keratin, elastin, and resilin proteins) for tissue regeneration and repair. Peptides and proteins as Biomaterials for Tissue Regeneration and Repair, pp. 175–204.

Coutu, A., Whitelaw, G., Roux, P., & Sealy, J. (2016). Earliest evidence for the ivory trade in southern Africa: Isotropic and ZooMS analysis of seventh-tenth century AD ivory from KwaZulu- Natal. *African Archeological Review, 33*, 411–435.

Department of Agriculture, Forestry and Fisheries. (2015). Sisal Production guideline. The Republic of South Africa.

Egelyng, H., Bosselmann, A. S., Warui, M., Maina, F., Mburu, J., & Gyau, A. (2017). Origin products from African forest: A Kenyan pathway to prosperity and green inclusive growth. *Forest Policy and Economics, 84*, 38–46.

FAO. (1977). The rubber tree. *FAO Economic and Social Development Series: Better Farming Series, 25*(3). ISBN 92-5-100156-1.

FAO Committee on commodity problems. (2017). Review of the sisal market industry: market prospects and policy. Tanga, United Republic of Tanzania 15–17 November 2017. CCP:HFJU 17/2.

FAO. (2021). Sisal. Future of fibers. https://www.fao.org/economic/futurefibres/fibres/sisal/en/.

Feinberg, H. M., & Johnson, M. (1982). The west African ivory trade during the eighteenth century: the "...and ivory" complex. *The International Journal of African Historical Studies, 15*(3), 435–453.

Fenske, J. (2014). Trees, tenure and conflict: rubber in colonial Benin. *Journal of Development Economics, 110*, 13.

Hartemink, A. E., & Wienk, J. F. (1995). Sisal production and soil fertility decline in Tanzania. *Outlook on Agriculture, 24*(2), 18–25.

Hashima, W. A., & Elhawary, I. A. (2021). The globalization of the Egyptian cotton spinning industry via engineering units. Part 2: The impact of the latest generation of Egyptian cotton on the quality factor of its yarn. *Alexandria Engineering Journal.* https://doi.org/10.1016/j.aej.2021.09.052.

Kayode, A. P. P., Nou9t, M. J. R., Linnemann, A. R., Hounhouigan, J. D., Berghofer, E., & Siebenhandi-Ehn, S. (2011). Uncommonly high levels of 3-Deoxyanthocyanidins and antioxidant capacity in the leaf sheaths of dye sorghum. *Journal of Agriculture and Food Chemistry, 59*(4), 1178–1184.

Kebede, A. T., Raina, S. K., & Kabaru, J. M. (2014). Structure, composition, and properties of silk from the African Wild Silkmoth, Anaphe panda (Boisduval) (Lepidoptera: Thaumetopoeidae). *International Journal of Insect Science, 6*, 9–14.

Konan, J. A., Koffi, K. K., & Zoro, A. I. B. (2018). Lignin biosynthesis rate is responsible for the varietal difference in fruit rind and seed coat hardness in the bottle gourd Lagenaria siceraria (Milina) Standley. *South African Journal of Botany, 117*, 276–281.

Kriger, C. E. (2005). Mapping the history of cotton textile production in precolonial West Africa. *African Economic History, 33*, 87–116.

Lamboni, L., Gauthier, M., Yang, G., & Wang, Q. (2015). Silk sericin: A versatile material for tissue engineering and drug delivery. *Biotechnology Advances, 33*(8), 1855–1867.

Lu, X., Sun, Y., Han, M., Chen, D., Wang, A., & Sun, K. (2022). Silk fibroin double-layer microneedles for the encapsulation and controlled release of triptorelin. *International Journal of Pharmaceutics, 613*, 121433.

Lima, P. R. L., Santos, R. J., Ferreira, S. R., & Filho, T. R. D. (2014). Characterization and treatment of sisal fiber residues for cement-based composite application. *Engenharia Agricola, 34*(5), 812–827.

McKinney, E., & Eicher, J. B. (2009). Unexpected luxury: Wild silk textile production among the Yoruba of Nigeria. *Textile: Cloth and Culture, 7*(1), 40–55.

Monson, J. (1993). From commerce to colonization: A history of the rubber trade in the Kilombero Valley of Tanzania, 1890–1914. *African Economic History, 21*, 113–130.

Moreira, P. I., Esteves, C., Palma, S. I. C. J., Ramou, E., Carvalho, A. L. M., & Roque, A. C. A. (2022). Synergy between silk fibroin and ionic liquids for active gas-sensing materials. *Materials Today Bio, 15*, 100290.

Munder, F., Furil, C., & Hempel, H. (2008). Advanced decortication technology for unretted bast fibers. *Journal of Natural Fibers, 1*(1), 49–65.

Negm, M., & Sanad, S. (2020). Cotton fibers, picking, ginning, spinning and weaving. Handbook of Natural fibers (2nd Ed). Volume 2: Processing and Applications. The Textile Institute Book Series, pp. 3–48.

Olatunji, O. (2018). Processing and modification of gum arabic in specific applications. In: Gum Arabic: Structure, Properties, Application, and Economics. Marios, A. A. (Ed.), Elselvier Accademic Press. UK, pp. 127–129. ISBN: 978-0-12-812002-6.

Olsen, K. M., & Schaal, B. A. (1999). Evidence on the origin of cassava: Phylogeography of Manihot esculenta. *Proceedings of National Academy of Science USA, 96*, 5586–5591.

Omosini, O. (1979). The rubber export trade in Ibadan, 1893–1904: Colonial innovation of rubber economy. *Journal of the Historical Society of Nigeria, 10*(1), 21–46.

Otim-Nape, G. W., Bua, A., Ssemakula, G., Acola, M. G., Baguma, Y., Ogawal, S., & Van der Grift, R. (2005). Cassava development in Uganda: A country case study towards a global cassava development strategy. In: A review of cassava in Africa with country case studies on Nigeria, Ghana, the United Republic of Tanzania, Uganda, and Benin. Proceedings of the Validation Forum on The Global Cassava Development Strategy. FAO, vol. 2.

Peek. (2004). Basketry: Africa in: African Folklore An Encyclopedia. Peek, P. & Yankah, K. (Eds.), Routledge. ISBN 9780415803724, pp. 20–22.

Peigler, R. S. (2020). Wild silks: Their entomological aspects and their textile applications. In: Handbook of Natural Fibres (2nd ed) Vol 1: Types, properties, and factors affecting breeding and cultivation. Woodhead Publishing Series in Textiles, pp. 715–745.

Phologolo, T., Yu, C., Mwasiagi, J. I., Maya, N., & LI, Z. F. (2012). Production and characterization of Kenyan sisal. *Asian Journal of Textile, 2*(2), 17–25.

Raina, M. A., Gloy, Y. S., & Gries, T. (2015). Weaving technologies for manufacturing denim. Denim: Manufacture, Finishing and Applications. Woodhead Publishing Series in Textiles, pp. 159–187.

Rath, T., Marti, O., Steyrer, B., Seidler, K., Addison, R., Holzhausen, E., & Stampfl, J. (2021). Developing an ivory-like material for stereolithography-based additive manufacturing. *Applied Materials Today, 23*, 101016.

Reddy, N. (2020). Silk: Materials, processes, and applications. A volume in The Textile Institute Book Series. Woodhead Publishing. ISBN 978-0-12-818495-0.

Savastano, Jr H., Santos, S. F., & Agopyan, V. (2009). Sustainability of vegetable fibers in construction. Sustainability of Construction Materials. Woodhead Publishing Series in Civil and Structural Engineering, pp. 55–81.

Slegtenhorst and Venter. (2009). An evaluation of interpreted technical and aesthetical design suitability in garments (aimed at a western market) in which Kalahari Tussah silk was used. 12th National DesignEducation Forum Conference Proceedings. Design Education Forum, Southern Africa, pp. 128–138.

Shamte S (2000) Overview of the sisal and henequen industry: a producers' perspective. Common Fund for Commodities- Alternative Applications for Sisal and Henequen-Technical Paper, No. 14. Proceedings of the Seminar held by the Food and Agriculture Organization of the UN (FAO) and the Common Fund for Commodities (CFC) held in Rome December 2000.

Spring, C., & Hudson, J. (2002). *Silk in Africa.* British Museum Press. ISBN: 9780714125633.

Steiner, M., Gould, A. R., Clark, T. J., & Burns, R. (2003). Induced Elephant (Loxodonta africana) tusk removal. *Journal of Zoo and Wildlife Medicine, 34*(1), 93–95.

Tageldin, M. A., Gomaa, A. A., & Hegazy, E. A. M. (2017). Respiratory symptoms and pulmonary function among cotton textile workers at Misr Company for spinning and weaving El-Mahalla Egypt. *Egyptian Journal of Chest Diseases and Tuberculosis, 66*(2), 369–376.

Tavuyanago, B., Mutami, N., & Mbenene, K. (2010). Traditional grain crops in pre-colonial and colonial Zimbabwe: A factor for food security and social cohesion among the Shona people. *Journal of Sustainable Development in Africa, 12*(6), 1–8.

Wadley, L., Trower, G., Backwell, L., & d'Errico, F. (2015). Traditional glue, adhesive and poison used for composite weapons by Ju/'hoan San in Nyae Nyae, Namibia. Implications for the evolution of hunting equipment in prehistory. *PLOSone, 10*(10), e0140269.

Wang, W., Huang, M. H., Dong, X. L., Chai, C. L., Pan, C. X., Tang, H., Chen, H. P., Dai, F. Y., Pam, M. H., & Lu, C. (2014). Combined effect of cameo2 and CBP on the cellular uptake of lutein in the silkworm, bombyx mori. *Plos/one, 9*(1), e86594.

Welsch, B., Jacobson, E. R., Kollias, G. V., Kramer, L., Gardner, H., & Page, D. C. (1989). Tusk extraction in the African elephant (Loxodonta africana). *Journal of Zoo and Wildlife Medicine, 20*(4), 446–453.

Yeboah, S. O., Hilger, T. H., & Kroschel, J. (2003). Thaumatococcus daniellii (Benn.) Benth. - a natural sweetener from the rainforest zone in West Africa with potential for income generation in small-scale farming.

Yu, C. (2015). *Natural textile fibers: Vegetable fibers* (pp. 29–56). Materials, Design and Technology. Woodhead Publishing Series on Textiles.

Zang, Y., Hu, Y., Xu, C., Wu, S., Wang, Y., Ning, Z., Han, Z., Si, Z., Shen, W., Zhang, Y., Fang, L., & Zhang, T. (2021). GhUBX controlling helical growth results in the production of stronger fiber. *iScience. 24*(8), 102930.

Zhang, Y., & Dong, H. (2020). Yield and fiber quality of cotton. Reference Module in Materials Science and Materials Engineering. Encyclopedia of Renewable and Sustainable Materials. Elsevier. ISBN 9780128131961, pp. 356–364.

Chapter 4
Plastic and Polymer Consumption in Africa Today

Abstract This chapter will explore the common types of polymer products available in the African markets today. The applications ranging from food to packaging, fashion, and healthcare will be explored. This is done with the aim of providing an understanding of the existing demand and trends in the plastic and polymer industry in the region. As plastics and polymers are used widely in the production of many consumer goods, the chapter also discusses consumer culture and its origin in the continent. This is linked to the consumption of plastic and polymer products in the region. Towards the end, the chapter also reports on some observational studies carried out.

Keywords Plastics · Consumerism · Polyethylene · Urbanization · Polystyrene · Polypropylene

Plastic Use and Rise of Consumerism in Africa

An Accenture report stated an estimated consumer spending in sub-Saharan Africa of 600 billion USD in 2010 and projected to a trillion USD by 2020. 51% of this comes from South Africa and Nigeria (Hatch et al., 2011), with an estimated population of 1.340598 billion according to the UN World population prospects for 2020. About 25% of this billion-plus population are in the middle class of the economy, and that number is expected to be twice as high by 2050. The projected growth for the African economy is two to three times that of the global economy (Hatch et al., 2011). Poverty in sub-Saharan Africa is also said to be decreasing, meaning more of the population will have disposable income for consumer spending. Reports also show increased urbanization with 60% of the population expected to be living in an urban environment by 2050. For example, the most populated African country, Nigeria has an estimated 52% urbanization rate (Criticaleye, 2010). This likely means increased use of building materials to construct homes, offices, and infrastructures. Already as of 2012, more than 50% of the African population own a mobile phone. This means more access to the consumer markets and information on products. For these reasons, the African region has become an attractive market for consumers. In addition to this, Africa is reported to have a growing middle class that is likely to spend more on consumer goods. A lot of these consumer goods are either made with or packaged with plastics.

Using the example of the Shoprite holdings, a well-known supermarket chain in various countries in Africa and the largest FMCG retailer in Africa. Shoprite opened its first store in 1979 and today has over 2913 stores in different countries in Africa including Botswana, Nigeria, and Mozambique. Its expansion into the rest of Africa began by opening its first store in Namibia in 1990, by which time it had around 71 stores in operation. In 1991, Shoprite obtained checkers and expanded to 241 stores. In 1995, the first Shoprite in Central Africa was opened in Zambia in the city of Lusaka. In 1997, it acquired another 157 supermarkets and 146 furniture stores in Eswatini and in the same year it expanded to Mozambique. The Shoprite group expanded to Zimbabwe and Uganda in 2000, however operations ended in these countries in 2013 and 2021, respectively. Similarly, it entered Egypt in 2001, however the 7 stores were closed by 2006 due to restrictions. In the same year, it expanded to Malawi and further expanded to other parts of its home country South Africa. In 2002, it expanded to Madagascar and Mauritius and Tanzania. Operations in these countries are either ended or sold. By 2003 Shoprites have been opened in Angola and Accra. In 2005, the first Shoprite is opened in Nigeria in the city of Lagos (As of May 31, 2021, operations of Shoprite stores in Nigeria have been acquired by Ketron Investments). In 2012, Shoprite store is opened in the Democratic Republic of Congo. In 2018, it opens up in Kenya (Closed in 2021). Within these periods it continued expansion in the different countries across Africa.

In 2021, Shoprite holdings reported revenue from the sale of merchandise of 168.0 Billion Rands. In 2020, it reported a revenue of 156.9 billion Rands. This was a rise from 147.5 billion rands reported in 2019 (Shoprite Annual Report, 2020). In 2000, the reported revenue was 18.430 billion Rands and in 1999 was 17.349 billion Rands (Shoprite Annual Report, 2000). These data have been presented to give an insight into the expansion of the fast-moving consumer goods within Africa, using the example of the largest FMCG retailer. Figure 4.1 shows an image taken in front of a Shoprite retail store at a mall in Lagos, Nigeria in February 2022. Some consumers can be seen exiting the store with the iconic yellow Shoprite plastic shopping bag.

There are several other micros to large supermarkets in Africa with similar growth ambitions and potential. Many well-known international brands are sold at these supermarkets as well as other formal and informal markets, kiosks, street hawking, and shops. For example, Nigeria is the fastest growing market for drinks brands such as Guinness. Sales in Nigeria exceed that of some western countries like Ireland. There are 27 Nestle factories across the African continent employing over 11,500 people. The value of plastic and polymers imported to the African continent annually runs into tens of billions of dollars. For example, between 1990 and 2017, plastic and polymer importation into Africa was valued at 285 billion USD (Babayemi et al., 2019). Considering that only about 9% of plastics get recycled globally, this means that a huge portion of this value is lost after each plastic gets used.

Taking Lagos, Nigeria as a case study, an observation of five different supermarkets was carried out and on average 10 out of 10 product types counted on the shelves were either made from plastics or packaged in plastic. The observation was the same

Fig. 4.1 In an image taken outside a Shoprite store in Lagos, Nigeria, some customers are seen exiting the store with the iconic yellow Shoprite plastic shopping bag, January 2022

for food, electronics, soft drinks, and household sections. Even in the wine and alcoholic beverages where glass bottles are used, plastic parts such as screw tops and inner caps were made of plastics. While some tinned products such as tomatoes and milk were made of only metals, some tin food products had plastic binders and some even plastic caps that allow the user to cover and store the product after reopening the seal. Picture frames made primarily out of wood and glass had transparent plastic film covering and cardboard cereal boxes had plastic as primary packaging inside to protect the product. Almost every product encountered at the supermarkets was made of plastics of one form or the other.

Consumerism in Africa is on the rise and from the indications, plastics make up a huge chunk of the products being consumed. In the following sections, the chapter looks at the origin of these plastics and polymers globally and in Africa. The chapter then takes a closer look at the types of plastics and polymers to consume and the rates at which they are being consumed in African countries relative to the rest of the world.

Tracing the Origin of Modern Polymers and Plastics in Africa

In the previous chapters, we have looked at the polymer products that were produced and traded in the early industry of Africa. All the plastics in the early industries were sourced from plant and animal sources. Some were traded as sourced while others required some additional processing and treatments. This section attempts to trace the earliest use of modern plastics in the African continent.

Although polymer-based materials such as cotton, rubber, and leather have been processed and traded for Millennials, the chemistry of polymers was developed only in 1920. The origin of modern polymers and plastics can be traced to 1926–1956 in Germany where Hermann Staudinger (Furukawa, 1998) carried out scientific work that formed the basis for polymer chemistry and materials science today. He established the theory of the polymer structure, fibers, plastics, and biological macromolecules. In 1953, he received the Nobel Prize for his work in Chemistry (American Chemical Society, 1999). He used natural rubber as the first polymer to demonstrate the chemical structure of polymers, demonstrating that rubber was indeed made up of repeating units of polyisoprene.

Rubber was one of the earliest commercial polymers. Initially, natural rubber was used in the production of erasers and some fabrics. Following the discovery of the vulcanization process in 1839 by Goodyear, there was a significant rise in the vulcanized rubber industry. By 1851 vulcanized rubber (ebonite) was patented. Cellulose derivatives were also commercialized in the early polymer industry. Cellulose nitrate was discovered in 1838 by John Wesley Hyatt (White, 1999) and commercialized by 1870. Plasticization of cellulose nitrate with camphor formed celluloid, a more processible material that found vast applications in the early industry. In 1865, cellulose acetate was discovered and this led to the production of materials like acetate rayon fibers and plastics of cellulose acetate (Billmeyer, 1981). Later it was discovered that cellulose could be dissolved in a suitable solvent to form viscose rayon and cellophane. These were used if fabrics and adhesive tapes. Thus, the early modern plastic and polymer industry mainly involved chemical or physical modification of naturally occurring polymers, mainly rubber and cellulose.

Bakelite can be said to be the first truly synthetic polymer to be commercialized. It is a thermosetting phenol–formaldehyde resin synthesized from a condensation reaction between a phenol and formaldehyde. It was named after the American chemist Leo Hendrik Baekeland who discovered it in 1908. By the 1920s vinyl, chloride-acetate resins were being produced on a large scale. The 1930s saw commercial production of polystyrene synthetic rubbers beginning in Germany and then in the United Stated from 1937 (Bilmeyer, 1981). In 1935, Carothers discovered nylon 66 while working at the DuPont lab. By 1938 a patent was granted for nylon and large-scale manufacturing had commenced in the US in Delaware. By the middle of 1939, women's hosiery made out of nylon fabrics was selling out (American Chemical Society, 1995). Nylon was used for the production of stockings due to its similarity to silk. It was also used in the production of toothbrush bristles albeit in much smaller quantities.

The decades that followed saw the introduction of several other forms of synthetic polymers introduced into the markets. These included polyvinyl chloride, poly-methyl methacrylate, silicones, ABS, polyethylene, polypropylene, and polycarbon-ates among many others. In 1965, 15 million tonnes of polymers were produced by 1996 and this figure rose to 150 million tonnes. This comprised 100 million tonnes of thermoplastics, 22 million tonnes of thermosets, 10 million tonnes of elastomers, and 18 million tonnes were synthetic fibers (American Chemical Society, 1999). From this information, we see that by the mid-twentieth-century plastics and other modern-day synthetic polymers were already being produced at a commercial scale in the countries like Germany and the United States. To understand the origins of modern-day plastics and polymers in the African continent, the trade between Africa and other regions is reviewed in the following paragraphs.

The colonial era resulted in the homogenization of African economies to some extent. The focus shifted from ecologically diverse produce to focus on cash crops and raw materials that were of interest to the international markets. In exchange for the exported raw materials and agricultural produce, Europeans imported manufactured goods which included firearms, alcohol, and textiles among others (Akyeampong, 2015). This access to manufactured goods and limiting of production within the African countries resulted in further dependency on imports. Generations were born not knowing any other products but those that were imported. The introduction of firearms made activities such as the slave trade more lucrative than agriculture or craftsmanship.

According to the literature on Africa in the 1300s, the Savannah to the South of the Sahel served as a source of cereals and fish as well as livestock which were sold for meat and leather across the Sahara. Rice was grown in the middle Niger with the River Niger acting as a source of water. The forest also served as a source of other products such as kola nuts. Written records from the 1500s show that cotton and silk produced within Africa were being worn in Tanzania. Up to the eighteenth century, imported textiles were only used as a status symbol, necessary clothing needs were met through local production. In great Zimbabwe, long-distance trade involved trading gold and ivory for Indian textiles and beads. In 1869, diamond was discovered at Kimberly in South Africa. In the same period, Europe and Asia got a direct sailing link through the opening of the Suez Canal in Egypt. They reduced the journey time between Europe and Asia.

In 1852, steamship service was introduced to the African waters. This allowed individuals who did not own ships to pay for the service which reduced the capital needed to trade in Africa. Traders and commission agents brought in goods. Between 1884 and 1885 came the Berlin conference where the African continent was divided up into colonies. Much of which was governed by Britain and France who were the main power at the time. The main goal as shown in various texts was to obtain raw materials from the colonies and in exchange import the manufactured goods from Europe. Local manufacturing was discouraged and so was trading with other colonies. Production in each colony was limited to only a few or a single product for export. For example, Kenya produced tea, Tanzania produced sisal, Malawi and Northern Nigeria produced cotton, Senegal produced groundnut, and Zambia produced copper. New currencies were introduced and a new wage-based work and taxes culture became the norm in African colonies. Cultivation of essential food crops was abandoned for the production of cash crops such as cotton.

The discovery of large crude oil reserves in Africa was also a contributor to its adoption of the modern plastic economy. For example, 70% of India's oil supply comes from Africa (Akyeampong, 2015). India in turn exports around 35 billion USD worth of products much of which are made of or packaged with plastics. The main raw material for plastic production remains hydrocarbons of fossil crude oil, a non-renewable resource. An estimated 12 million barrels of oil go into the production of the nearly 100 billion plastic bags produced every year (Behuria, 2021). It is expected that based on the current consumption rate, 20% of the fossil oil produced by 2050 will go towards the production of plastics (UN, 2018). Libya, Nigeria, Algeria, Angola, and Sudan combinedly produce around 90% of the crude oil in the continent. 38 countries out of 53 are net oil importers (African Development Bank, 2009). In 1957, oil production in the African region only accounted for around 0.3% of global production volume. By 1967 8 to 9% of global oil production was coming from Africa at around 279.5 million metric tonnes (Baker, 2008). By 2021 the crude oil reserves in Africa were reported at 125.3 billion barrels producing around 6.9 million barrels daily (Statista, 2021a, b, c). In 2020, Nigeria lead as the top oil producer in the continent with 86.9 million tonnes produced 2020 (Statista, 2021a, b, c). With a large natural resource that is also a source of plastics and polymers. The reliance of the economy on crude oil export with little manufacturing provides an incentive for importing convenient consumer goods which modern living has become so dependent on. Much of these contain fossil-based plastics and polymers.

Following World War II, international organizations such as the United Nations, the World Bank, the International Trade Organization, and the International Monetary Funds were found. By the time of independence between the 1950 and the 1960s, the taste and consumption patterns of Africans have become more modernized. This meant that demands for a standard quality of life for Africans required modern infrastructures such as electricity, modern buildings, and roads and there is a huge cost of such required foreign investments. Attempts to industrialize Africa post-independent included the formation of regional economic integration bodies such as the East African Community which was set up in 1967, the Economic Community of West African States set up in 1975, and the Southern African Development Coordination Conference (SADCC) set up in 1980. The African economy declined in 1980 and the 1990s saw an increase in foreign investments from Asia and Arab countries in Africa. By 2009 China had become Africa's leading trade partner and lender. By 2015 China-Africa trade was valued at 135.9 billion USD. That of India was 35 billion USD by 2008 (Frynas & Paulo, 2007). This trade typically includes textiles, equipment, machinery, pharmaceutics, instruments, telecommunications, and information technologies, in exchange for raw materials.

From the above, it can be said that prior to the invention of synthetic plastics in 1908 and the widespread commercialization of plastics by 1939, trade between

Africa and the Western World had been well established. There was a movement of goods between African countries and the rest of the world. Therefore, it is expected that as soon as these synthetic plastics and polymer products became available in the market in Europe and America, they would have also been exported to the rest of the world and exchanged for raw materials.

Classification of Consumer Plastics

Other than the resin identification number, plastics can also be classified by their end-use into commodity plastics, engineering plastics, and high-performance plastics. High-performance plastics are those with exceptional properties such as high-temperature resistance and excellent physical properties in extreme conditions. Examples are polyethylene ether ketone (PEEK) and polytetrafluoroethylene (PTFE) (Verma et al., 2021). Engineering plastics generally have good mechanical properties and can be used for load-bearing applications. Examples are polycarbonate and nylon.

Much of the concern with littering has been largely from commodity plastics. These are plastics that are commonly found in everyday products and are more frequently used. For example, 56 million metric tonnes of polypropylene were produced globally in 2018 with this figure projected to reach 88 million metric tonnes by 2026 (GlobeNewswire, 2019a, b). Production of high-performance plastics like PTFE, for example, is still in the thousands of tonnes. In 2015, global production of PTFE was at 165 thousand metric tonnes and is projected to be 247 thousand metric tonnes by 2022 (GlobeNewswire, 2019a, b). More commonly, these high-performance plastics are used for more long-term applications. For example, PEEK is used in surgical implants and prosthetics. Commodity plastics tend to be used for shorter term applications such as food packaging and shopping bags. Engineering plastics like nylon used in, for example, fishing nets and rope in sailing often end up in the ocean as litter.

Some common applications of different plastic products as observed from plastic products on display at supermarkets in Lagos Nigeria are shown in Table 4.1. The range of plastic products found in the markets and supermarkets in Lagos, Nigeria is also present in the list of products that were listed in a publication by WEF (2016) for global plastic consumption. This similarity can be attributed to globalization and the widening of the international supply chain.

Plastic and Polymer Consumption Rates

Globally, 8300 million tonnes of plastics are reported to have been produced since the early period of plastics' introduction to the market to 2015 (Geyer et al., 2017). This figure is expected to grow further in the next decade or two (UNEP, 2018). Given a

Table 4.1 Common plastic types, their resin identification number (RIN), and their applications as observed in Lagos, Nigeria

Plastic type	RIN	Example applications
PET Polyethylene terephthalate	1	Water, soft drinks bottle, salad domes, fruit punnets, peanut butter containers, salad dressing containers
HDPE High-density polyethylene	2	Milk bottles, freezer bags, juice bottles, ice cream containers, shampoo bottles, detergent bottles, chemicals bottles
PVC Polyvinyl chloride	3	Cling wrap, cosmetics container
LDPE Low-density polyethylene	4	Shrink wraps, bin bags, squeeze bottles
PP Polypropylene	5	Ice cream containers, chips bags, microwave dishes
PS Polystyrene EPS Expanded polystyrene	6	CD cases, party cups, plastic cutlery, glassware substitute Foams PS coffee cups, takeaway clamshell packs, meat trays, packaging foam
Others Examples: PC (Polycarbonate) Nylon PMMA (Polymethylmethacrylate) ABS (Acrylonitrile butadiene styrene)	7	Refillable water dispenser bottles Hair nets, fishing nets, ropes Vehicle lightings, prosthodontics Machined prototypes, pipe, toys

global population of 7.75 billion people alive and 66.7% of whom are aged between 20 and 100 years of age weigh on average 62 kg. This would bring the adult human population's total weight to an estimated 321.9 million tonnes. What this implies is that humanity creates the equivalent of its own weight in plastic waste annually. Accumulation of plastic waste in the environment has begun to have a devastating impact on the ecosystem, the list of which is the unpleasant sights of plastic waste floating on the ocean, clogging drainage, and littering the streets. Aquatic life gets harmed by choking on plastics, and plastics blocking sunlight for photosynthesizing water plants which are crucial to maintaining balance in the environment and serving as food for other organisms. On land, there are reports of livestock choking on plastic leading to their death and economic loss to the owner.

Plastic is being produced more than almost any other material in the world. An estimated 367 million tonnes of plastics were produced in 2020 (Statista, 2021a, b, c). The annual production rate of plastic since 2015 is estimated at 400 million tonnes. In 1950, an estimated 1.5 million tonnes were produced (UNEP, 2018). Thus indicating that the rate of plastic production since its introduction is exponentially increasing. Of this, around 79% accumulates as waste in the landfills and environment, 12% get incinerated while around 9% have been recycled. With recent efforts and increased environmental consciousness of the public, manufacturers, and government, these figures are gradually changing. Data from WEF indicates that 14% of plastics were recycled (albeit only 2% were effectively recycled, 8% downcycled into lower value products, and about 4% lost in the recycling process) (WEF, 2016). This is an increase from the average recycling rate of 9%. Here we explore the rate of plastic consumption in Africa.

Plastic production in Africa accounts for less than 1% of global single-use plastic production. In one of the first studies to analyze plastic consumption in the African continent (Babayemi et al., 2019), it was reported that between the years 1990 and 2017, 86.14 million tonnes of primary polymers and 31.5 million tonnes of plastic products were imported into the African continent. Extrapolating this to the 54 African countries with a population of 1.216 billion people, this means an estimated 172 million tonnes of plastics and other polymers imported into the continent over the 10-year period. Other products with plastic components also contribute significantly to the number of plastics. This pushes up the number to around 230 million tonnes of plastics being imported into the continent within the 10-year period analyzed.

According to the UNCTAD report (Barrowclough et al., 2020), Morocco makes the list as the 45th top importer of plastic packaging in 2018. Nigeria and Ghana appear as the 10th and 12th importers of plastic fishing nets in the world. Nigeria, Kenya, Morocco, and Egypt are among the top importers of intermediate manufactured goods from China as of 2018.

Some of the plastics consumed in Africa are also produced within the continents. These producers contribute an estimated 15 Mt between 2009 and 2015 (Babayemi et al., 2019). Many of the plastic products and packaging found in the supermarkets in this study were shipped from countries like China, the UK, the US, Germany, France, Spain, and other African countries.

To get a firsthand picture of the rate of plastic product consumption in a part of Lagos, Nigeria observations were carried out at selected supermarkets within Lagos. On two visits to the supermarkets, the number of shopping bags being taken out by customers was counted over a period of 10 minutes. Where it is uncertain if bags were doubled, it was counted as one. Therefore, the figures are more likely to be underestimated. In this observation, on average 64 plastic bags left the supermarket within 10 min. This observation was carried out on a Saturday afternoon between 1 and 2 pm. This period can be described as a peak period as most people are likely to do their shopping on the weekend. Nonetheless, this gives an idea of how much of this category of plastics is being consumed. Assuming that this average applies best to the busiest period between 12 and 3 pm on Saturdays, then this peak period on Saturdays contributes an estimated 59,904 plastic bags to the total annual plastic bags taken out from this particular store.

Supermarket chains such as Shoprite offer consumers the option of purchasing reusable shopping bags. Figure 4.2 shows an example of a supermarket shelves displaying various designs of reusable shopping bags made from cotton, polypropylene, and other materials. On another visit to five different supermarkets in Lagos, Nigeria, the products on the shelves were observed for plastic content. Table 4.2 summarizes some of the different types of plastic products, and the type of plastics used.

Fig. 4.2 Reusable shopping bags are available for purchase at some supermarkets in Lagos, April 2022

	Product	Plastic type
Table 4.2 Some sample plastic products found in five different supermarkets in Lagos between December 2021 and January 2022	Plastic bottled water	PET
	Rechargeable fan casing	PVC
	Party cups	PS
	Shampoo bottles	HDPE PP PET
	Ice cream containers	PP
	Refillable water dispenser bottles	PC

Conclusion

Plastic consumption in the region is rising, and as a result urbanization is increasing which is accompanied by increased dependence on plastic products. The relatively low financial cost of plastics compared to other alternatives means more people opt for plastics where it is available. With increased globalization, plastic products from all parts of the world are imported into African countries. Alternatives like woven shopping bags to replace flimsy polyethylene shopping bags are available.

References

African Development Bank. (2009). *Oil and gas in Africa*. Oxford University Press, pp. 1-2. ISBN 978-0-19-956578-8.

Akyeampong, E. (2015). *History of African trade*. African Export-Import Bank. ISN 978-92-95097-16-2

American Chemical Society. (1995). *A national historic chemical landmark: The first nylon plant Dupont Seaford Delaware*.

American Chemical Society. (1999). *An international historic chemical landmark: The foundation of polymer science by Hermann Staudinger (1881–1965)*. Freiburg, Baden-Wurttemberg.

Babayemi, O. J., Nnorom, I. C., Osibanjo, O., & Weber, R. (2019). Ensuring sustainability in plastics used in Africa: Consumption, waste generation, and projections. *Environmental Sciences Europe, 31*, 60–79.

Baker, J. (2008). Oil and African development. *The Journal of Modern African Studies, 15*(2), 175–212.

Barrowclough, D., Birkbeck, C. D., & Christen, J. (2020). Global trade in plastics: insights from the first life-cycle trade database. United Nations Conference on Trade and Development. UNCTAD/SER.RP/2020/12. Research Paper No 53.

Behuria, P. (2021). Ban the (plastic) bag? Explaining the variation in the implementation of plastic bag bans in Rwanda, Kenya, and Uganda. *Politics and Space*, 1–18.

Billmeyer, J. R. F. W. (1981). *Textbook of Polymer Science*. Wiley-Interscience.

Frynas, J. G., & Paulo, M. (2007). A new scramble for African oil? Historical, political, and business perspectives. *African Affairs, 106*(423), 229–251.

Furukawa, Y. (1998). *Inventing polymer science: Staudinger, Carothers and the emergence of macromolecular chemistry*. University of Pennsylvania Press.

Geyer, R., Jambeck, J. R., & Law, K. L. (2017). Production, use, and the fate of all plastics ever made. *Science Advances, 3*(7), 1–5. https://doi.org/10.1126/sciadv.1700782

GlobeNewswire (2019a) Global polytetrafluoroethylene (PTFE) market analysis, trends, and forecast, 2016–2018 & 2019–2024 with focus on the United States industry. Retrieved January 7, 2022 from https://www.globenewswire.com/news-release/2019a/06/12/1867377/0/en/global-polyttetrafluoroethylene-PTFE-market-analysis-trends-and-forecasts-2016-2018-2019-2024-with-focus-on-the-United-States-industry.html

GlobeNewswire. (2019b) Polypropylene market to reach USD 155.57 billion by 2026|Reports and data. Retrieved January 7, 2022 from https://www.globenewswire.com/news-release/2019/08/01/1895698/0/en/polypropylene-market-to-reach-USD-155-57-billion-by-2026-reports-and-data.html.

Hatch, G., Becker, P., & van Zyl, M. (2011). The dynamic African consumer market: exploring growth opportunities in sub-Saharan Africa. Accenture. 11-0457_LL/7-1842.

Shoprite Holdings LTD. (2000). Annual Report.

Shoprite Holdings LTD. (2020). Integrated Annual Report.

Statista (2021a) Global plastic production 1950–2020. Retrieved January 7, 2022 from https://www.statista.com/statistics/282732/global-production-of-plastics-since-1950.

Statista (2021b) Main oil-producing countries in Africa 2020. Retrieved January 21, 2021b.

Statista (2021c) Proved crude oil reserves in Africa 2010–2021c. Retrieved January 8, 2021c from https:www.statista.com/statistics/1178138/crude-oil-reserves-in-africa/.

UNEP. (2018). *Single-use plastics: A roadmap for sustainability* (Rev. ed., vol 6): p 4

Verma, S., Sharma, N., Kango, S., & Sharma, S. (2021). Developments of PEEK (polyetheretherketone) is a biomedical material: A focused review. *European Polymer Journal, 147*, 110295.

White, J. L. (1999). Fifth of a series: Pioneer of polymer processing John Wesley Hyatt (1837–1920). *International Polymer Processing, 14*(4), 314–314.

World Economic Forum. (2016). *The new plastics economy*. Rethinking the future of plastics. www.weforum.org/docs/WEF_The_New_Plastics_Economy.pdf

Chapter 5
Plastic Bans in Africa

Abstract More countries in Africa have introduced plastic bans as a measure to address the global plastic crisis. The impact of these bans varies across the region from Rwanda where it has been largely successful with measurable impacts to other regions where the implementation hasn't been so successful due to various factors like enforcement and lack of practical alternatives. This chapter takes a look at the plastic ban as an instrument of the government to stop plastic pollution with a focus on the African region.

Keywords Plastic ban · Policy · Reduction strategies · Plastic waste · Sustainable

Plastic Bans Across the World

Across the world, different countries have risen to one of the greatest challenges of the twenty-first century, plastic pollution. Measures, countries have adopted to address this problem, include the use of policy instruments in the form of bans and levies. Currently, Africa is leading in this approach owing to the success of Rwanda in banning single-use plastics, and the impact this has had on its environment and economy (Behuria, 2021). The first country-wide ban on single-use plastics was in Bangladesh in 2002 (Chowdhury et al., 2021). Prior to this at the state level, there was a ban in the state of Sikkim, India in 1998. This ban was imposed on plastic bags and wrappers. Recent years have seen around 60 countries across the globe introducing single-use plastic bans (UNEP, 2018). These bans are typically on specified types of single-use plastics. Other than bans, governments have at their disposal other instruments for addressing the plastic pollution problem. This is mainly in the form of levies and taxes on plastics (Jambeck et al., 2021). Other strategies for reducing plastic pollution include the promotion of eco-friendly alternatives, strengthening the waste management system of the country, facilitating reduction strategies, agreements with private organizations, and generating public pressure.

The effectiveness of plastic bans and other policy instruments has been varied. According to a UN report focused on single-use plastic bans across the world (UNEP, 2018), of the countries where single-use plastic bans have been introduced, 30% report a significant drop in usage of plastic bags within a year of the ban while

20% of them report the bans having no impact on the use and littering of plastics. The rest of the countries could not provide data on the impact of the ban. This chapter takes a closer look at the single-use plastic ban in Africa, implementation strategies, impacts, and other issues relating to such bans. Thus far there is a global trend towards reducing the use of plastic. Bans and levies are increasingly being introduced at the local, national, or regional levels. In many cases, the ban applies to specific types of single-use plastics. Many countries have adopted the strategy of banning the most avoidable single-use plastics. The types of plastics ban vary for different locations (UNEP, 2018). As of 2018, 60 countries have introduced one form of single-use plastic ban or the other. Out of these, 33 are African countries. Observing the trends indicates that other policy instruments like levis and taxes are preferred by the governments of more industrialized countries like the UK. Asia, thus far, seems to have put in less effort in the introduction of plastic bans compared to other regions. China announced a ban on single-use plastic bags and straws in January 2021 following an initial ban on plastic utensils and tableware. On July 3, 2021, the EU introduced a single-use plastic ban that includes 10 of the most problematic single-use plastics. These are the plastics that are most commonly found littering the environment. The top 10 oceans polluting waste reported in 2020 are cigarette butts, plastic drinks bottles, packaged food wrappers, and other uncategorized trash such as face masks, plastic bottle caps, shopping bags, straws and stirrers, takeaway containers, beverage cans, glass drinks bottles, other plastic bags and styrofoam packages (International Coastal Cleanup Report, 2021). Other than the glass bottles, metal cans, and other nonplastic items that might fall within the "others" category, many of these top problematic wastes are plastics.

More levies than bans have been introduced in European countries. France as of January 1, 2022 introduced bans on plastic fruit and vegetable packaging such as punnets and plastic bags used in the packaging of fruits and vegetables (De Clercq, 2021). The ban applied to 30 fruits and vegetables such as oranges and cucumbers where plastic packaging is considered avoidable. In England, October 2020 saw the introduction of bans on some single-use plastic products; stirrers, cotton buds, and plastic straws (Department of Environment, Food and Rural Affairs, 2021). In August of the following year plans for bans on plastic cutleries, plates, cups, and other single-use plastic bans beginning with public consultation and a call for evidence were announced.

In the Oceania region, Papua New Guinea introduced bans on non-biodegradable plastic bags and in Australia, a ban was introduced on single-use plastic bags (UNEP, 2018). Most countries in South America with policies against single-use plastics have used levies rather than bans. Eight states in the United States have banned plastic bags, cups, and some selected single-use plastic products.

It is expected that coming years will see further bans and levies placed on single-use plastics as the world moves towards a more circular economy. This is unless technology evolves towards a more circular plastic economy. Developments in chemical recycling could offer a future of endlessly recycling plastics. This along with a leak-proof waste management system could present a case for lifting bans or removing the need for imposing them in the first place. Other chapters in this book discuss different plastic waste processing technologies and their development in Africa.

Leaders in Plastic Ban Across Africa

The East African region has made notable moves towards single-use plastic bans in recent years. The different countries give different perspectives on the implementation and impact of the plastic ban. The main countries in the region that have introduced plastic bans at the national level are Rwanda, Kenya, and Uganda.

Rwanda has placed itself as a leading pioneer in the regional and global plastic pollution challenge through its successful implementation of a national single-use plastic bag ban, specifically polyethylene bags that were less than 100 microns in thickness. This began in 2005 (Behuria, 2021) triggered by a prior The Rwandan Environmental Management Authority-funded research study revealed the extent of the impact of plastic pollution on the environment and economy (Kabenga & Musabe, 2003). The ban on the use and importation of single-use plastic bags was fully implemented into force by 2008. These efforts in banning single-use plastic bags contributed to the capital of Rwanda, Kigali, being nominated for the 2008 UN habitat award as one of the cleanest cities in Africa. Although the approach of the government to banning single-use plastics has been described as heavy-handed, following the ban, some businesses were forced to terminate their production processes within weeks with an unpleasant financial impact.

Since the introduction of plastic bans in 2005, Rwanda has seen a significant boost in its tourism industry. According to a study on tourism surveys and tourism satellite accounts, Rwanda's tourism sector contributed around 89,000 jobs for the country 261.2 billion Rwandan Francs from internal tourism consumption.

In its pursuit of a reputation as a global environmental leader, Rwanda is ambitious to extend the plastic bag ban towards eliminating all single-use plastics in the country. The Green Fund FONERWA has also been set up to fund environmental-related projects across Rwanda. In addition to this, Rwanda has also introduced an exceptionally tough emissions target of 16% reduction by 2030, and reforestation targets to increase the land area covered by forest to 30% by 2030.

Kenya introduced a ban on plastic bags earlier than Rwanda did however it faced several delays and was not implemented until 2018. Nonetheless, it has now successfully implemented bans on single-use plastic bags and shows the positive impact of the policy on its environment. Kenya also has a thriving tourism industry and does benefit from adopting an image of a pristine and green beautiful country. However, when compared to Rwanda, Kenya also has a relatively stronger and more organized manufacturing sector. The manufacturers in Kenya have the Kenyan Association of Manufacturers (KAM), which has played an active role in preventing the implementation of plastic bans in the country. During the period where the plastic bag ban was delayed, there was a growth in the annual production rate of plastic bags. There were protests from businesses that produced and traded plastic bags. The feared impact was the loss of employment for means of earning that the plastic bag usage provided these stakeholders. Despite the business power of manufacturing associations and opposition to plastic bans in Kenya, the ban was finally implemented in 2018 and plastic bag manufacturing has been forced to seize.

Despite Kenya having a more active manufacturing sector, the government's pursuit of a plastic ban is attributed to the desire to remain competitive in the tourism industry. Part of the strategies that Kenya has used in promoting its international image as an environmentally conscious country includes the successful lobbying to secure the location of the united nation environmental protection agency in its capital, Nairobi. Kenya has a long-standing reputation as a green nation with active environmental movements. For example, the Wildlife Club Movement of Kenya has been in existence since 1968. The Wild Life Club of Kenya is the first of its kind to have ever been formed in the world. The 2004 Nobel Peace Prize was awarded to Wangari Maathai, the founder of the Green Belt Movement and a former Assistant Minister of Environment.

The Kenyan service sector contributes more to its GDP than the manufacturing sector and tourism contributes a huge chunk to the service sector. The service sector contributes around half of the GDP while the manufacturing sector contributes around 13%. In 2014, tourism contributed around 10.5% to the GDP (Njoya & Seetaram, 2018). Tourism earning from international arrivals for Kenya in 2019 was 164 billion Kenyan Shillings, which is equivalent to around 1.6 billion USD (Statista, 2021). This was prior to the disruption caused by recent insecurities and the Pandemic which resulted in the figure dropping to 37 billion Kenyan Shillings in 2020.

Bans on single-use plastic bags have been announced in Uganda on different occasions; however, they have not been well implemented or enforced. From all indications, the adoption of plastic bag bans in Rwanda is largely motivated by the need to adopt an image of a greener tourist destination to attract foreign visitors. Uganda on the other hand since the discovery of large oil reserves in 2006 has been more focused on setting the country up as an oil-producing nation (Mawejje, 2019). It, therefore, has little motivation to cultivate the image of a green, pristine tourist destination. The production of oil even before its commencement is already harming the wildlife in Uganda. The location of one of the oil wells will inevitably disrupt the wildlife and ecosystem of the Albertine region, which is regarded as one of the most biodiverse regions in the world and a place where large numbers of mountain gorillas are found (Isabieye, 2020). It should be noted that although the oil was struck in 2006, the journey to oil production in Uganda had begun in the 1920s when assessments pointed out the Albertine region as a potential oil-rich site. This followed further assessments of other areas and licensing of sites for oil exploration midsts major events such as the Second World War and other regional events, which delayed the process (Langer et al, 2020). Therefore, the investments and commitment of the individuals and organizations in the discovery of oil in Uganda undoubtedly pose a challenge to prioritizing tourism over oil.

A joint press release by the National Environment Management Authority and Uganda National Bureau of Standards on December 7, 2021 (NEMA & UNBS, 2021) announced the enforcement of the bans on plastic bags below 30 microns thickness.

It reported that of the 47 factories that have been inspected, 21 had their operations suspended due to non-compliant. The resumed effort on the plastic bag ban might be aided by the delays in the commencement of oil production in the country since the discovery in 2006 due to several factors. The lack of anticipated progress in the oil sector gives way for the government to consider not abandoning its tourism sector by investing in environmental conservation activities such as banning plastic bags.

Botswana introduced a levy that resulted in a significant reduction in the use of plastic bags by as much as 50% within 18 months. There are discussions to proceed to ban because, despite the reduced use of plastic as a result of the levies, this did not translate to reduced plastic pollution. In Cameroon's ban on nonbiodegradable plastic ban, despite resulting in the smuggling of plastic bags from neighboring countries, the government provides incentives such as monetary rewards to citizens for plastic waste collection. In 2015, this resulted in around 100 thousand kilograms of plastic waste being collected.

In Chad, there are reports of a visible reduction in plastic littering the environment following the introduction of the plastic bag ban. Eritrea has also successfully introduced plastic bag bans, which have resulted in the reduction in the plastic bags blocking drainage and other associated problems. Morocco is another country in Africa that has reported a success story from the plastic bag ban. The ban was introduced in 2009 with little success in implementation; however, the government improved on its implementation efforts in 2016 (Alami, 2016) leading to the seizure of 421 tonnes of plastic bags and today use of plastic bags is almost completely eradicated from the country.

Plastic bag bans in Nigeria, Côte d'Ivoire, Egypt, Somalia, Senegal, South Africa, Tanzania, Tunisia, Ethiopia, Gambia, Guinea Bissau, Kenya, Malawi, Mali, Mauritania, Mauritius, Zimbabwe, Benini, Burkina Faso, Cape Verde and Niger (Nwafor & Walker, 2020) have either being ineffective or there isn't sufficient information on the effectiveness of their implementation. Either Way, proper documentation and monitoring of the implementation and compliance are important in ensuring that the ban is successful. Therefore, a lack of information suggests that the policy isn't doing very well.

Factors Influencing Successful Plastic Ban

Plastic bans have been vastly motivated by studies and reports that highlight the adverse impact that plastic pollution has had in the given country or globally. For example in Rwanda and Kenya, plastic pollution had the adverse impact of blocked drainage systems, soil pollution, and ingestion by livestock, which caused fatal damage. All of these also had direct and indirect economic impacts. Governments, prior to introducing a ban hold consultations with experts to plan out appropriate measures and if a ban or levy can be effective in addressing the problem. Ideally, stakeholders are engaged in dialogues on the potential impact of the intended ban and strategies for implementation. A decision is then taken based on facts presented to introduce a ban and/or a levy on certain types of plastics, usually single-use plastics. The plastics banned are ideally the most problematic as well as most practical to ban. A final decision is then made and the ban is introduced. This can be at a local state level, country, or region. A period is given between informing the public of the intention to ban and the actual implementation and enforcement of the ban. For example, France announced its intention to ban plastic use in the packaging of some fruits and vegetables on October 11, 2021 (De Clercq, 2021), the ban was put into force on January 1, 2022. This allows a period of adjustment for all stakeholders making compliance more possible.

Some form of punishment is often attached to a ban, this can be a fine or in some cases imprisonment. Kenya's anti-plastic law is regarded as the most severe so far. Businesses or individuals found violating the anti-plastic law are liable to a fine of up to 40,000 USD or up to 4 years in prison. When compared to, for example, the United States, in New York, first-time offenders are fined 250 USD, second-time offenders 500 USD, and multiple offenders are fined the maximum of 1000 USD. In China, offenders are fined between 1,545 and 15,460 USD. In Rwanda, plastic bag smugglers are liable to 6 months imprisonment if caught. The government agents also carry out unscheduled inspections of businesses to ensure compliance (Behuria, 2021).

The United Nations in its 2018 report, which assessed the progress and lessons learned from single-use plastic bans across the world, provides a 10-step roadmap to guide governments and stakeholders towards successful implementation of single-use plastic bans. The recommendations include beginning with the most problematic plastics and those that are dispensable, consulting the general public and other stakeholders prior to implementation, assessing the alternatives to bans and levies, predictive analysis of expected impact, raising public awareness, educating the general public, suppliers, and manufacturers on alternatives to the single-use plastics that are to be banned, providing incentives and assistance to businesses to ease compliance, responsibly reinvesting the revenues from fines, transparency and enforcement, and effective monitoring and inspection strategies.

Bans on single-use plastics can be overturned after they are introduced. This will mean wasted resources that have been put into implementing them in the first place. For example, the plastic ban in New York initially faced opposition from recyclers and manufacturers, and other stakeholders. This resulted in the lifting of the ban. However, in 2017, the ban was later re-enacted after further investigations revealed that the recycling of styrofoam was not viable both economically and environmentally. There are existing laws in states like Arizona, Michigan, and Missouri against banning single-use plastics. The justification for such moves is the severe economic hardship this might present to the plastic and recycling industries. Therefore, a ban can result in a lot of legal battles and additional administrative work.

The plastic and polymer industry is very important. It is a significant contributor to GDP, it provides employment, and plastics and polymer products are important in international trade. Plastic packaging is largely single-use make up around 50% of plastic production globally (UN, 2018). Plastic manufacturers can exert business power that may significantly hinder the success of the implementation of a single-use plastics ban or levies. To put into context, within Africa, Rwanda which has been able to implement a single-use plastic ban did so amidst hardly any active plastic manufacturing in the country. In the more industrialized region, the United States, however, has over 881 plastics and resin manufacturing companies as of 2021. In Europe 5,000 plastics and rubber, companies are listed in the business wire directory. The world's largest plastic manufacturer, China, has until recently hardly any bans on plastic use or production. Looking at the global production figures (UN, 2018), 21% of single-use plastic is produced in North America, northeast Asia produces 26%, Europe produces 16%, and Africa produces only 1%.

The existing socio-economic realities of the country or region are also a significant factor to consider when implementing plastic bans. Taking the case of Kibera slums in Nairobi Kenya, in the poorest parts of the country where adequate toilet facilities were not available; there were reports in 2000 of individuals resorting to defecating in plastic bags (Lusambili, 2011). These are then disposed of in water drainage, open spaces, or even rooftops. Humans will naturally find ways to address a need with what is within reach. The use of single-use plastic in such a manner would have not been necessary if this basic need for public toilet facilities had been met. Some single-use plastic products provide access to facilities that are not available in deprived areas of some countries. The sachet water is an example of such. Water filled and sealed in polyethylene water sachets are for some the only way to access clean drinking water in many cities and slums in some parts of Africa. Therefore in such a case, a ban on single-use plastics without making adequate affordable and accessible solutions would cause severe hardship.

In some places, bans were announced officially; however, there was little effort to enforce them. It is important to carry out inspections and enforce the law once it is introduced. New York, for example, tasked three agencies of the government to carry out annual inspections or 311 investigations of affected establishments. The DSNY provides the public with information on suitable alternatives through its website. Training and educational materials are provided to educate businesses towards compliance.

In some cases, the use of the banned plastics is unavoidable. In such cases, waivers and exemptions are necessary. For example, the US makes provisions in the form of hardship waivers for small businesses that earn below a gross income of $500,000. Such waivers apply to businesses that apply and provide evidence that seizing the use of the banned packaging will cause the business financial hardship. Rwanda also provides some special dispensation that allows manufacturers to use plastic packaging in their products (Behuria, 2021). To do such, the manufacturer must obtain permission from the government to use or produce nonbiodegradable plastic packaging.

Rwanda, since the 1990s, has charted its development towards a service-based economy. In recent decades, Rwanda's service sector has outperformed its primary and secondary economic sectors (agriculture and manufacturing and construction sectors). Between 1994 and 2019, the service-based sector grew from < 30% of GDP to over 50% of GDP. It is also recognized as the fastest-growing economy in Africa.

At the heart of its service sector is tourism. Tourism pulls around 374 million USD as of 2017 with this targeted to grow to 800 million USD by 2024.

The government directing its effort towards developing the service sector is attributed to the geographical positioning of Rwanda. It is a land-locked country that makes the importation of materials and machinery for manufacturing and export of manufactured goods relatively more challenging. The government, therefore, seeks to grow the economy through tourism. It is much easier to attract visitors to travel into and within the country for tourism and conferences than it is to move goods and machinery across the borders. Tourism is significantly impacted by its international reputation hence focusing on building its reputation as a pristine and green inviting destination country helps boosts its tourism.

Harnessing the Promise of Plastics Last Forever

The ultimate goal of a ban on plastics is to reduce the accumulation of plastic waste in the environment. However, if not well thought out or well implemented, a ban can end up having the opposite or another adverse impact. Thus far, an estimated 9 billion tonnes of plastics have been produced since the 1950s, and in recent years, around 400 million tonnes of plastics are produced annually with only around 9–12% of this getting recycled. What do we do with all the plastics that are still out there? One way is to recover them, effectively recycle them, and re-use them. So long as the demand exists for a product, this motivates innovation in the use of the product. So while a ban on all single-use plastics might lead to a reduction in plastic waste accumulating in the environment, it might seter efforts to recover and recycle existing plastics. One example of such an effort is Coca-Cola's prototype plastic bottle produced from plastics recovered from the marine (Lingle, 2019). Before now, effectively recycling plastics back into food and beverage products has proven challenging due to the high purity required for plastics to be used in such applications. With developments in recycling technology, the prospects of discarded plastics being a feedstock for producing new food-grade plastics seems more possible. This gives the possibility of endlessly recyclable products and hence a circular plastic economy.

PETCO in South Africa is an example of a company that presents an alternative to the plastic ban in Africa (http://petco.co.za). A company focused on the recycling of PET plastic waste, it has managed to develop a market for rPET within the country and outside and also created jobs and entrepreneurship opportunities.

Despite the negative press that plastics seem to have received, the possibility of endlessly recyclable plastics could mean that a ban on plastics might deny modern civilization from reaping the true promise of plastics; plastics last forever. If plastics can be endlessly recycled, this will remove the pressure from other resources such as metal and wood. The mining of metals eventually takes a toll on the environment. Similarly cutting down trees and using natural cellulosic resources for the production of paper and cardboard which often are not even as durable as so-called single-use plastics eventually strain the natural rate of renewal. With these in mind, a ban on single-use plastics might be counterproductive or disadvantageous.

Voluntary Plastic Reduction Strategies

Supermarket stores, companies, and individuals in some parts of Africa have taken measures to reduce the use of plastics without the need for bans. In South Africa, a model was introduced for encouraging the use of refillable containers instead of single-use plastics for purchasing goods called the Gcwalisa means refill in the Zulu language. Dispensers are provided to the goods traders that allow them to sell variable quantities of goods to customers. The customers purchase the goods that can vary from foodstuff to homecare products, in reusable containers. This began in 2019 as a pilot project in Johannesburg. The ability to purchase variable volumes reduces product wastage and the need for plastic packages for pre-packed fixed quantities. The product is metered into the containers using an electronically controlled system implemented into the dispensers. This system was mainly targeted at the reduction of plastic waste generated from the informal spaza shops across South Africa. It was developed through a partnership between two companies, Smollan and DYDX (https://smollan.com). The equivalent for the formal sector called the Smart fill dispenser has also been developed by the partnership.

Impact of Plastic Bans in Africa

Placing a nationwide ban and levy on plastics is a recent approach that only dates back to 2002. In the UN report on single-use plastic bans and levies across the world, several countries did not have enough information on which to assess the impact of the introduced ban or levy. There are of course the obvious advantages of eliminating plastic waste and problems associated with plastic waste. Where the ban or levy has been successfully implemented and data collected on its impact a ban on single-use plastics has shown other added advantages.

Through the single-use plastic ban, Rwanda has adopted the image of a pristine and beautiful country that is a pleasure to visit. This branding has contributed to the growth of Rwanda's tourism-powered economy. On a global scale, countries across the globe collectively implementing plastic bans promote international cooperation and a globally shared vision of a plastic pollution-free world.

During the period of the assessment by the UN, Eritrea achieved a significant drop in plastic bags blocking drains and waterways, and landfills. In Morocco, plastic bags use decreased to almost zero in the country. Some countries like Ethiopia haven't reported many benefits from introducing plastic bans. This is attributed to unclear enforcement and in, for example, Cameroon, without feasible and affordable alternatives, smuggling of plastic bags becomes lucrative. There are reports of smuggling and the black market for plastic bags in The Congo. A ban can lead to hardship for lower-income people if affordable and feasible alternatives are not provided or subsidized.

While the technology exists to mechanically and chemically recycle plastics exist. The logistics and the process of recycling consume, time, manpower, materials, and energy. All of which are limited resources. Advanced recycling technologies such as feedstock chemical recycling and pyrolysis are still at relatively early stages for scale-up. Bans and levies aimed at reducing the rate at which plastic waste is generated reduce the pressure on these limited resources and prevent the existing facilities for recycling from being overwhelmed. Already plastic pollution cost to the global aquatic ecosystem is estimated at an annual cost of 13 billion dollars.

Restricting the use of problematic plastics through bans and levies can help drive demand towards more sustainable alternative products. This demand for more sustainable options invariably drives innovation in sustainable alternatives. The fines and levies generated can also be used to fund innovative sustainable projects and research and improve the waste management system. In this regard, levies are better suited to generating funds for such public projects since full compliance with a ban generates no fines.

Despite the bad reputation single-use plastics have gained, when life cycle assessment and overall impacts are compared, turns out, that single-use plastics aren't the absolute worst option after all. The alternative materials for the same applications in some cases prove to have a worse impact when compared. For example, reusable plastic bags made out of materials like cotton or polypropylene are only a better alternative to single-use plastic bags if they are re-used a minimum of 50 times before they are discarded for recycling. If not they have a worse impact. While single-use plastic bags can be reused several times with moderate care, not tear or puncture, paper bags due to their flimsy nature are mostly only good for a single-use and are not waterproof. Another popular alternative to plastic bags is cotton bags. Where the energy, water, and land resources for the production of cotton fabric are considered, despite its biodegradability, cotton cost more to the environment than single-use plastics. Cotton bags are only a better alternative to recycled single-use plastic bags if they are used around 1000 times.

Countries in Africa can learn some lessons from Japan and its approach to plastic use and waste management system. Japan remains one of the countries that hasn't imposed any anti-plastic policies, instead it has a leak-proof waste management system that ensures effective recycling of plastic waste.

Conclusion

So would a single-use plastics ban work for Africa? Well, it depends. From Kigali to New York City, data from across the world suggest that yes, bans do work but only if they are well suited, adequately implemented, and effectively enforced. Therefore when considering plastic bans either regionally, nationally, or locally, the issues discussed in this chapter should be considered for the specific case.

References

Alami, A. (2016). Going green: Morrocco bans use of plastic bags. Aljazeera. Retrieved 18 January, 2022, from https://www.aljazeera.com/news/2016/7/1/going-green-morocco-bans-use-of-plastic-bags.

Behuria, P. (2021), Ban the (plastic) bag? Explaining the variation in the implementation of plastic bag bans in Rwanda, Kenya, and Uganda. *Politics and Space, 0*(0), 1–18.

Chowdhury, G. W., Koldewey, H. J., Duncan, E., Napper, I. E., Niloy, H. M. N., Nelms, S. E., Sarker, S., Bhola, S., & Nishat, B. (2021). Plastic pollution in aquatic systems in A review of current knowledge. *Science of the Total Environment, 761*, 143285.

De Clercq, G. V. (2021). New law in France will save 1 billion pieces of single-use plastic annually. World Economic Forum. https://www.weforum.org/agenda/2021/10/how-france-plans-to-significantly-reduce-its-plastic-waste-from-2022/.

Department for Environment, Food & Rural Affairs and The Rt Hon George Eustice MP. (2021). Plans unvieled to ban single-use plastics. Press Release 20 November 2021. Retrieved 13 January, 2022, from https://www.gov.uk/government/news/plans-inveiled-to-ban-single-use-plastics.

Isabieye, M. (2020). Environmental sustainability: An afterthought or key objective for Uganda's oil sector? Oil wealth and development in Uganda and Beyond: Prospects, Opportunities, and Challenges. (Langer A, Ukiwo U, Mbadazi P (eds). Leuven University Press, pp. 226–238.

Jambeck, J., Hardesty, B. D., Brooks, A. L., Friend, T., Teleki, K., Fabres, J., Beaudoin, Y., Bamba, A., Francis, J., Ribbink, A. J., Baleta, T., Bouwman, H., Knox, J., & Wilcox, C. (2021). Challenges and emerging solutions to the land-based plastic waste issue in Africa. *Marine Policy., 96*, 256–263.

Kabenga, P., & Musabe, T. (2003). *Etude sur la gestion des déchets plastiques au Rwanda*. University of Rwanda.

Langer, A., Ukiwo, U., & Mbabazi, P. (2020). Oil wealth and development in Uganda and beyond: prospects, opportunities, and challenges. Leuven University Press.

Lingle, R. (2019). First of a kind Coca-Cola PET bottles made from ocean plastics. Plasticstoday: Packaging. October 22 2019. Retrieved 10 January, 2022, from https://www.plasticstoday.com/packaging/first-kind-coca-cola-pet-bottles-made-ocean-plastics.

Lusambili, A. (2011). "it is our dirty little secret": An ethnographic study of the flying toilets in Kibera slums, Nairobi. STEPS Centre, Economic and Social Research Council. ISBN 978 185864 976 5.

NEMA & UNBS. (2021). Press release: Enforcement of the ban on plastic carrier bags below 30 microns.

Njoya, E., & Seetaram, N. (2018). Tourism contribution to poverty alleviation in Kenya: A dynamic computable general equilibrium analysis. *Journal of Travel Research, 57*(4), 513–524.

Nwafor, N., & Walker, T. R. (2020). Plastic bag prohibition bill: A developing story of crass legalism aiming to reduce plastic marine pollution in Nigeria. *Marine Policy, 120*, 104160.

Mawejje, J. (2019). The oil discovery in Uganda's Albertine region: Local expectations, involvement, and impact. *The Extractive Industries and Society, 6*(1), 129–135.

Reuters. (2020). Kenya's tourism earnings, arrivals in 2019. Reuters.

Statista. (2021). Tourism earnings from international arrivals in Kenya 2005–2020. Retrieved 17 January, 2022, from https://www.statista.com/statistics/1140037/tourism-earnings-from-international-arrival-in-kenya/.

UNEP. (2018). SINGLE-USE PLASTICS: A Roadmap for Sustainability (Rev. ed., pp. vi; 6), pp. 4.

Chapter 6
Plastics Recycling in Africa

Abstract Recycling plays a key role in sustainable plastic waste management and the adoption of a circular plastic economy. This chapter will review the plastic recycling technologies used, scale, new developments, and future perspectives on plastic recycling in Africa. The discussions are based on academic literature, available data in the public domain, and observations from physical visits to some parts of the region. This chapter looks specifically at mechanical recycling; another chapter in this book is dedicated to chemical plastic recycling.

Keywords Recycling · Mechanical recycling · Upcycling · rPET · Recycling infrastructure

Plastic Recycling

Plastic recycling rates globally are around 14%. Considering that the recycling rate of for example steel and aluminum is around 50% (OECD, 2018), the recycling rate of plastics is relatively low. Although this is the global average recycling rate, the rate of recycling varies significantly by country and region. Recycling rates in Europe are around 30% (Geyer et al., 2017) while there are insufficient data on the recycling rates in Africa. Generally, the rate of recycling is increasing in recent years (Hoornweg and (Bhada-tata, 2012). The informal sector is estimated to be achieved around 20–40% recycling in some cities in African countries (Wilson et al., 2009).

Some plastic types are more recycled than others. For example, around 10% of PET and HDPE get recycled while PS and PP hardly get recycled. This is partly due to the fact that the former are less prone to lose their properties upon recycling while the latter are more likely to have their properties altered after the first couple of recycling processes. PS carries the risk of releasing hazardous chemicals like benzene in the process of recycling while PP tends to become more brittle when it is recycled. Plastic waste collection and recycling has become a global business with plastic wastes being traded internationally through extensive networks (Hoornweg & Bhada-Tata, 2012).

Recycling is the process of reforming plastic products back into the same or another plastic product that is either of the same, lower, or higher quality of the

© The Author(s), under exclusive license to Springer Nature Singapore Pte Ltd. 2022   73
O. Olatunji, *Plastic and Polymer Industry by Region*,
https://doi.org/10.1007/978-981-19-5231-9_6

application. There are different forms of recycling. These are classified either based on the method used to recycle or the end use of the resulting product from recycling. The focus of this chapter is the mechanical recycling of thermoplastics. In all recycling's first stage is the collection of the waste plastics and taking them to the point of recycling which could be a factory or recycling depot. Once the plastics get to the recycling depot further sorting is often required. This can be automated sorting which requires machinery and labor. After the sorting, the plastics get cleaned. In some models, the plastic waste is simply compressed into bales and shipped off to the buyer. In other cases, the recycling companies carry out further stages of recycling to add value to the recycled plastic. Here, the cleaned plastics then get shredded and broken down into smaller particles. This is typically done in a granulator. Depending on the level of purity required, further sorting can be carried out using processes such as floatation, air blowing, or more advanced sorting using near-infrared (Wu et al., 2020). Heat and mechanical shearing then melt and reform the plastic into smaller pellets in a pelletizer. The plastic pellets are then sold to plastic manufactured or processed in-house into end products.

Plastics break down when exposed to sunlight and/or heat. This might take around 450 years during which the plastics may break down into microplastics and even nanoplastics. The additives and colors used in plastics tend to include heavy metals and other toxins. Some plastics have UV stabilizers, which delay the degradation by sunlight. Therefore, landfilling or burning of plastics has a much higher accumulation rate than the degradation rates with plastics being dumped daily. Recycling helps keep the plastic away from landfill and the environment for longer. Since the invention of synthetic plastics, several types of plastics have been developed and are in use today. Each plastic type has its own properties and attributes and not all of them can be recycled. Plastics are often assigned resin identification numbers from 1 to 7 that are usually inscribed on the plastic products. The inscription can be the resin identification number written in the middle of the three chasing arrow symbols with or without the abbreviated plastic typewritten under the arrows. An example is shown in Fig. 6.1.

These are PET (polyethylene terephthalate), HDPE (high-density polyethylene), PVC (polyvinyl chloride), LDPE (low-density polyethylene), PP (polypropylene) and PS (polystyrene). Plastic number 7 refers to all other plastics such as nylons, ABS, polycarbonates, and many more. For example, single-use shopping bags are mostly made using HDPE, woven shopping bags are made using polypropylene, plastic drink bottles are most commonly made with PET, and takeaway bowls are most commonly made from PP while the foamed ones are made from PS. While the chasing arrow sign is taken to indicate that a plastic product or package is recyclable, it doesn't mean it gets recycled. Some plastic types are not well suited for many recycling plants. PS for example is prone to releasing toxins upon recycling, which makes its recycling hazardous. PP tends to become more brittle when recycled and LDPE plastic bags are difficult to process and handle for recycling plants. Most plastic's properties deteriorate upon mechanical recycling with further deterioration with subsequent cycles. Often virgin plastics must be added to recycled plastics to attain acceptable properties.

Fig. 6.1 Example of the resin identification number inscribed on a strawberry package made from rPET assigned the resin identification number 1

Stains on plastic glues used, labels attached, printed inks, and other contaminants all result in less efficient recycling. A colored plastic bottle poses more difficulty in recycling compared to a clear one, for example. The technology to remove ink from waste plastic packaging has been developed (Fullana & Lozano, 2015); however, this additional process means more resource input and time in the recycling process. One approach is to rethink and redesign products to use less of this additional material, which makes recycling more difficult and reduces the quality of the recycled plastic. For example, newer PET bottles seen in the markets today are designed with labels that only make use of one line of glue that sticks less to the bottle and is easier to peel off. This is an improvement to labels that use more glue on a larger surface that are more difficult to remove.

Plastics such as polystyrene and PVC release toxins into the environment during recycling. Thus, recycling these plastics comes with serious health hazards. It is therefore important that manufacturers consider the product's end of life throughout the design and production stages. In recent years, more companies are paying more attention to the environmental impact and sustainability of their products. For example, Coca-Cola now uses 100% recycled PET bottles in its products across 30 markets.

The collection and sorting stage is often overlooked as technically uncomplicated compared to the other stages of recycling. However, the collection and sorting of plastic prove rather challenging on a large scale. The moment a used plastic gets out into the environment, getting it to the recycling station to some extent is left to chance, particularly, where an effective waste management system is not in place. The piece of plastic can get thrown on the streets, clogging a drainage system, fall into a lagoon, get washed up on a beach, left drifting across the ocean, choke a sea turtle, or get blown around by the wind across a street in the city. Collection of the recyclable plastic waste from the source reduces the chances of the plastic getting to pollute the environment. Plastic waste tracking technologies have been introduced to address the collection challenge. For example, the Dallas Sanitation Service App has been developed to serve as a digital platform that facilitates plastic waste collection in bulk. Through the app, members of the community register and receive alerts on recycling collection dates.

More producers are now inclined to increase the recycled plastic content of their products. This increased demand for more recycled content can be attributed to the rise in the price of rPET which as of 2021 rPET has risen to a dollar per pound according to the wall street journal report (https://www.wsj.com/articles/empty-plastic-bottles-go-from-trash-to-hot-commodity-11636455644). Virgin PET is priced at just a fraction of a dollar that virgin PET costs. This emerging scarcity is not due to the unavailability of used plastics, there are millions of tonnes floating around in the oceans with around 8 million tonnes being added every year. However, this widely available plastic resource is not easily accessible. Getting the discarded plastics requires resource input for collection and sorting. Further advanced sorting technologies are required to sort the plastic further in order to achieve food-grade purity.

The fourth industrial revolution has seen the development of tools that if optimally utilized can greatly help lead the world towards a circular plastic economy. Tools of the fourth industrial revolution like near-infrared spectroscopy allow efficient automated sorting to high purity in plastic recycling. The future cost of cleaning up the plastic waste that is going to be released in the coming years assuming the current rate of accumulation persists, far exceeds the cost of implementing reduction strategies to prevent the accumulation of mismanaged plastic waste in the environment. While recycling alone does not hold the solution to the plastic pollution problem, it is a key strategy to ensure plastics remain in the supply loop for the optimal time period.

The global commitment report jointly produced by the Ellen MacArthur Foundation and the United Nations states that although not enough to achieve a fully circular plastic economy, plastic recycling has been the key driver of plastic waste management in past years. With the world waking up to the dangers of plastic pollution the use of virgin plastics in products is expected to drop faster from 2025 onwards (Ellen MacArthur Foundation & UN, 2021) based on the commitments and targets that have been set by various countries and organizations.

The consumption of virgin plastic is expected to reduce significantly in the coming years. Businesses that are signatory to the global commitment have demonstrated this commitment through a 1.2% drop in the number of virgin plastics that were used in the products manufactured and traded by these companies. This is a remarkable change because from 1950 to 2015 production of virgin plastics increased from an estimated 2 million metric tonnes to 300 million metric tonnes. Since 2019, over a thousand organizations including businesses, government organizations, and others, have committed to a joint vision of moving the world towards a circular plastic economy. This is one where the idea of a used plastic as waste is replaced with one where the used plastic can be reused or become a resource for creating other valuable products.

The third industrial revolution saw the invention of what has been described as a miracle material, plastics. Used in almost every aspect of modern life, plastics have become so ubiquitous, that they have become an environmental nuisance in many parts of the world. The great challenge of the twenty-first century is to do away with the throw-away culture of the past era and move towards a more circular plastic economy. Recycling is only one of the means to achieving a circular plastic economy. It is one of the three Rs of sustainability; reduce, reuse, and recycle. Reducing the number of plastics produced annually and ensuring that these plastic products can be reused several times also contribute significantly towards addressing the global plastic pollution challenge.

Every year approximately 300 million tonnes of plastics become discarded as waste (UNEP, 2018). If plastic consumption persists at the current rate, without recycling, it is estimated that there will be 12 billion tonnes of plastics littering the aquatic and terrestrial environment by 2050. That much plastic in the environment is bound to have a devastating impact. Choking of marine life, posing a threat to food security, and increased flooding as a result of plastic waste clogging drainage systems, are only some of the anticipated impacts of such levels of plastic pollution in the environment.

The problem is exacerbated when these plastics break down into microplastics. These can cause a whole host of known and even yet undiscovered damage to the ecosystem. These include abrasion on the digestive systems and internal organs of aquatic organisms that mistake them for food and eat them, zooplanktons consuming microplastics resulting in deoxygenation of the ocean (Kyale et al., 2021), they serve as a surface for harmful microbes to adhere to, and accumulate to hazardous levels in the system. At the very least, plastic recycling can reduce these adverse impacts by preventing plastic waste from ending up in the environment. Microplastics are covered in more detail in another chapter of this book.

Collection and Recycling Infrastructure

A large portion of the spending in plastic waste management in African countries goes towards collection while in more developed regions like Europe, more goes into disposal through processes like recycling and incineration. It is estimated that, of the plastics to have ever been produced, 30% are still in use (Geyer et al., 2017).

In African countries like Uganda, recycling options seem to be limited. Kampala, the capital city of Uganda for example generates 180 tonnes of plastic waste per day (Balcom et al., 2021). Around 40% of this plastic waste is collected and the collected waste is discarded in landfills in Kiteezi or destined for open burning. The plastics that don't get collected or burnt end up littering the roads, clogging drainages, or water bodies. Some of the plastic wastes end up contaminating the soil or getting ingested by livestock which may have a damaging impact on the livestock and the health of humans likely to consume the livestock (Priyanka & Dey, 2021).

From observations of daily life in Lagos, it is common practice to see plastic discarded along the roads, drainage, bushes, and waterways. Often these are done by pedestrians disposing of packages like sachet water that was just drunk, it is also a common sight to see plastic packages like PET bottles, food wrappers, and sachets thrown out of moving cars into the gutters or into the streets. However, there are several initiatives that have been implemented by both private and public institutions as well as individual efforts to curb plastic littering in the city. Figure 6.2 shows disposal facilities provided through public–private partnerships to curb littering in selected public places in Lagos.

The Inherent Value of Used Plastics

Based on the estimated amount of plastics produced annually and the percentage of this that is recycled, the recycled plastic production globally is around 46 million tonnes, which is an estimated 12% of plastics produced (Geyer et al., 2017). The rest is virgin plastics. Until 2017 when it banned imports of some waste plastics, China was the world's highest importer of waste plastics. Annual plastic exports to China were valued at 0.9 billion USD in 2016. Therefore, recycled plastic does contribute a decent share of the global plastic market.

Many of the plastics that are discarded still have a lot of value within them. Harnessing this value within the used plastic can yield an abundance of feedstock for producing sustainable plastics, chemicals, and energy. The technology to harness them also exists; some like mechanical recycling have been well established over the decades while others such as pyrolysis are still at the pilot scale. Globally, just around 9% of the plastics produced get recycled. This makes used plastics a largely underutilized resource since the various products and energy (Rehan et al., 2017) that can be obtained from the discarded plastics are not being harnessed.

Today PET is one of the most well-recycled single-use plastics. The technology to recycle PET has advanced to the stage where recycled PET or rPET is now used as food-grade packaging. As the global food chain expands and trade routes become more extended, these rPET packages have made it to stores in Lagos Nigeria. Figure 6.3 is an example of rPET grape fruit packaging that was obtained from a supermarket in Lagos, Nigeria.

(a)

(b)   (c)

Fig. 6.2 Plastic bottle bin provided at **a** the Johnson Jakande Tinubu park in Lagos, Nigeria **b** and **c** at the Ikeja City Mall in Lagos, Nigeria. These bins are provided and serviced as a public–private partnership between the state waste management authorities and private companies. June 2022

Fig. 6.3 A rPET grape fruit packaging observed at a supermarket in Lagos, February 2022

Other than the value inherent in plastic as a material, the recycling venture also creates value in other aspects. Jobs are created at different points of the recycling process from waste picking to factory work to distribution. Entrepreneurs can generate revenue from various recycling endeavors. In Ghana, for example, a UNFCCC report reveals how engaging in plastic waste collection has provided a source of income for some (UNFCCC, 2022). Across Africa, various startups have ventured into different aspects of recycling. Examples are Wecyclers in Nigeria and Mr. Green in Kenya.

Despite banning the importation of plastic waste in 2017 (Wen et al., 2021), China's previous interest in importing waste plastics from other countries is an indication of the real value of used plastics. Since the 1990s, China has accepted plastic waste from other countries and recycled them into other products, much of which are then exported. The later ban on the importation of plastic waste emerged as a result of increasing pressure on domestic waste management which results in over 70% of the imported plastic waste being discarded.

To successfully reach the goal of a circular plastic economy, a multifaceted approach must be adopted in waste management. Recycling should not be the only single waste management strategy. Rather recycling should be used to augment the other strategies for plastic waste management. A road map that allows countries to integrate multiple measures into plastic waste management is presented by The World Economic Forum (WEF, 2019). The road map includes using policy instruments like plastic bans, considering sustainability in all aspects of product development and design and directing resources towards the development of sustainable alternatives.

As part of the efforts to convert plastic waste into high-value products, new studies have shown that waste plastics can be converted into graphite. This has been demonstrated using waste PET bottles (Ko et al., 2020). In a pyrolysis process carried out at a temperature of 900 °C. Pyrolysis was followed by graphitization at 2400 °C over the boron catalyst. The graphene obtained showed high crystallinity and a relatively higher graphitization rate compared to similar processes. Graphite is used in high-end applications like barrey anodes in electric vehicles, energy storage devices and fillers in composites. The graphite can also be converted into graphene through processes such as microwave-assisted liquid phase exfoliation and dry exfoliation methods (Ma et al., 2020). Conversion of plastic waste into such a high-value product can help cover the cost of sourcing the waste and the additional processes required for recycling compared to production from the virgin material. This is particularly important as recent years have seen a significant rise in the price of post-consumer plastics. For example, the PCR in consumer electronics in 2020 was valued at 11.4 million USD and is expected to grow at a cumulative annual growth rate of 8.6% between 2021 and 2028 (Grandview Research, 2021).

Since conventional mechanical recycling requires energy input, the overall sustainability of the process depends on the source of energy used. The most feasible outcome is that the energy input is as low as possible and as sustainable as possible. Burning a lot of fossil fuel to recycle plastic waste may not be overall sustainable. It should be noted that the figures available on plastic pollution vary with the methods used to derive the figures. Some research studies using methods such as standard epifluorescence microscopy (Brandon et al., 2019), generalized linear model (Powell et al., 2016) and geographic information systems (GIS) analysis (Blanco et al., 2018) show that the amount of plastic waste in the environment is likely underestimated and could be more than official values presented. This underestimation could be due to factors like excluding microplastics or previously unavailable data in some areas. For example, in the US recent studies show that the 32.2 million tonnes of plastics estimated to be in landfills is actually 44 million tonnes (Milbrandt et al., 2022). In 2012 while the official figures reported 122 million tonnes of plastics in landfills in the US, a research study using generalized linear model of both active and closed landfills, estimated 262 million tonnes for the same year (Powell et al., 2016). Globally 4.5 to 10 billion USD can be saved by implementing more leak proof plastic recycling systems. Based on the previous variation in data there could be even more inherent value in the plastic waste than the official figures suggest.

Extended energy analysis of roof tiles used in Uganda made from plastic and sand mixture showed that a net 16,462 MJ per tonne of energy is saved when the roof tiles are used in asphalt road construction compared to other methods of disposal (Balcom et al., 2021). When the roof tiles are disposed of by pyrolysis a net 11,303 MJ per tonne was saved. The study compared pyrolysis, burying, landfilling, and incineration with mixing with asphalt in road construction. The energy consumed in transporting the raw materials, processing and transportation and energy consumed by CO_2 scrubbers for all cases. Despite burying and landfilling not requiring process energy inputs for example heating and pulverizing, the energy saved in the use of bitumen and cement in road construction makes this method of disposal more favorable in terms of energy. Therefore when considering solutions for plastic waste disposal, in depth analysis such as life cycle assessments and extended energy analysis is required to consider the best option for the country as a whole in terms of energy and environment.

Plastic Recycling Enterprises Across Africa

Every stage of recycling consumes energy, time, manpower, and other resources (Gopinath et al., 2020). Collection of the plastics from the point of generation requires some logistics. There are various approaches that have been used around the world and across Africa. In countries like Nigeria, Tanzania, Uganda, and Ghana, waste pickers are widely involved at this stage. Recycling businesses such as Recycle Point and Wecyclers operate models whereby plastic waste is collected using bicycles, push carts, or other means, and rewards are given to the collectors at the drop-off centers. At the 2021 Plast print Pack event in Lagos, a manufacturer showcased recycling machines that we're able to convert used PET bottles into new rPET bottles that are fit for use in packaging water again. Thus indicating that the technology for recycling plastics is available to recyclers in the African region through the global supply chain.

One of the key challenges in recycling plastics lies in the additional energy input that is needed to get a used plastic from being picked up at the source to a new pellet that can be used to produce the same or other product again. This can make producing a product from virgin plastic less costly and complex compared to making it from recycled plastic. Furthermore, the additional stages also mean more emissions in creating the recycled pallet. Plastics are projected to contribute 17% to global greenhouse emissions by 2050. The combined carbon dioxide emission from landfilling, recycling, and incineration of waste plastics in 2015 were an estimated 1.8 billion tonnes (Zheng & Suh, 2019).

This section explores various plastic recycling businesses across Africa. The list is by no means exhaustive, the purpose is to provide some selected examples of recycling businesses and their approach to recycling.

Mr Green Kenya

Mr Green Africa located in Nairobi Kenya has become the first B Corporation in Africa. It produces recycled PP, HDPE, PET of different colors and garages that it supplies to plastic product manufacturing companies. The waste plastics are sourced from waste collectors and operate on a fair trade basis to ensure that waste pickers are well compensated for their work. As of January 2022, it announces it engages 2500 waste collectors, directly employs 120 people and has so far recycled 6,000 tonnes of plastics. Mr Green is set up as a social enterprise that puts at its center the goal to positively impact informal plastic waste collectors by integrating them into the plastic recycling industry. Its investors include key players in the plastic industries and leading investors. These include DOW, Minderoo, AlphaMundi Group (AMG), The Bestseller Foundation, DOB Equity and the Global Innovation Fund. Although its main focus is on plastics, it also collects packaging cartons.

Wecyclers Nigeria

Wecyclers in Lagos Nigeria collect plastic bottles and bags as well as other waste materials like glass bottles and cans. These are then sold to larger plastic recycling companies. The difference here is that recyclers collect directly from consumers. These could be the plastics used in homes, estates or businesses that supply the pre-sorted plastics in exchange for incentives. These incentives include food items, call cards, and others. Using this model the collection and sorting are spread out at numerous small scales. The labor employed is diverse as it ranges from a person sorton at home to a business owner getting rid of plastic waste used by customers in for example a canteen. Another company in Nigeria, recycle points, operates on a similar model.

TakaTaka Kenya

TakaTaka in Kenya requires its suppliers to sort their waste into organic and inorganic. The organic waste gets converted into compost that is then sold to local farmers. The inorganic waste which includes plastics mostly gets sold to recycling companies. A fraction gets converted to other products for example glass bottles in tumblers. TakaTaka reports it collects around 10 tonnes of waste daily. This consists of plastic and other materials like glass and cans. It recycles up to 95% of the waste it receives

Proplast Senegal

Proplast focuses on the conversion of plastic waste into granules which it sells to plastic processing companies who then further process the granules into finished or semi-finished products. It employs 100 people and records a plastic recycling rate of 1500 tonnes annually. Conversion of plastic waste into granules involves shredding the sorted plastics into smaller sizes. This can then be processed into smaller particle sizes in a granulator or pulverizer. In this form, the plastics are easier to handle and meter into plastic processing equipment.

Polyco South Africa

Polyco takes a mixed approach to plastic waste management by setting up various projects that address plastic waste in different ways. One such project provides mobile kiosks located within informal settlements where people in the community can exchange plastic waste for cash and other incentives. The kiosk serves as an access to waste management facilities that would otherwise be unavailable to the communities in the informal settlement.

SoleRebels Ethiopia

Every year estimated 1 billion waste tires are discarded in the world causing an environmental nuisance. SoleRebels, a recycling company in Ethiopia, focuses on converting used tires into fashionable footwear. They convert the used tires into strips, which can then be woven on traditional Ethiopian hand looms. The project operates on fairtrade in order to pay the workers up to $4 \times$ the minimum wage in Ethiopia.

Ocean Sole, Kenya

Plastic flip-flops have become staple footwear in warm climates as is in many parts of Africa like Kenya. As they are used often, they eventually get worn out and are disposed of. Where they are not properly disposed of, they end up polluting the land and aquatic environment. Ocean Sole is dedicated to recovering these flip-flops from the environment and recycling them into art.

Repurpose Schoolbags, South Africa

This company collects discarded plastic bags and plastic billboards and converts them into school bags for underprivileged school children. The bags have in-built solar panels, which serve as a reading lamp for the purpose of aiding the children in studying and doing class assignments where public electricity supply is not available.

Ecoplast Kenya

Plastic pollution is only one of the environmental challenges the earth has to address. There is also the issue of deforestation among others. Deforestation is partly fueled by the demand for wood products such as lumber. Ecoplasts address these two key issues by recycling waste plastic and agricultural waste into plastic lumber. These have applications such as fencing and flooring. Its operation is supported by a waste management app. Ecoplast collects pre-sorted wastes from individuals and businesses and in return gives them points (Jambeck et al., 2018).

From the businesses discussed above, we see that there is a range of approaches, recycling businesses have taken towards plastic recycling. Most businesses choose a specific aspect of the recycling process and apply an innovation that makes the process more effective and beneficial to their target niche. Many of these businesses can be classified as social enterprises.

The different stages of recycling add value to plastics. For example, plastic waste that has been sorted into a single type of plastic is valued higher than a pile of different types of plastics and other materials. The labor and time that goes into the sorting come as a cost. Similarly shredded and granulated plastics have more value than the unprocessed used plastic in their original product forms.

Plastic Recycling Initiatives by Multinational Companies in Africa

Initiatives such as CLIP, African Marine Waste Network, and The African Circular Economy Alliance exist to address various aspects of the plastic pollution challenge at international, regional, and national levels (Sadan & de Kock, 2020). The focus of these organizations ranges from preventing plastics from getting into the marine environment, recommending best practices, and monitoring the plastic litter in the environment. Nigeria, Tanzania, and Egypt are the two African countries among the top 10 mismanaged plastic waste-generating countries in the world. Nigeria is at number 6 generating an estimated 1.90 million tonnes in 2015. Tanzania is right below Nigeria at number 7 with 1.77 million tonnes of plastic waste generated in the same year. Egypt is at number 10 with an estimated 1.6 million tonnes generated in 2015 (Lebreton & Andrady, 2019). This section reviews some of the plastic recycling initiatives some multinational companies have set up in these countries.

SA Plastics Pact

Also in South Africa, The SA Plastics Pact was introduced in January 2020. It is part of the global network of Ellen MacArthur Foundation's Plastic Pact where members commit to a list of terms specified in the pact. Members include Coca-Cola, Tuffy, Unilever, Polyoak, HomeChoice, and Unilever amongst others (www.saplasticpact. org.za). The four main commitments are that by 2025 the country aims to redesign, innovate and rebuild alternative packaging that eliminates the need for plastic, all plastic packaging must be reusable, recyclable, or compostable, and 70% of plastic packaging should be effectively recycled and to have all plastic packaging contain on average 30% post-consumer recycled content (Sadan & de Kock, 2020).

ReFlexNG

The DOW Company launched a pilot-scale recycling project in Lagos State, Nigeria on July 14, 2020. The goal of the project is to collect and recycle plastics across Lagos. This is part of its global commitment titled "Stop the waste" through which it has set the target of having 1 million tonnes of plastics collected, reused, or recycled by 2030. This initiative is in partnership with Nigerian-based Recycle Points, Omnik, and Lagos Business School Sustainability Centre. This project focuses on post-consumer plastic water sachets which are a common source of affordable drinking water in Nigeria.

RecyclePoints (www.recyclepoints.com) handles the collection of the plastic water sachet through its collection kiosks and app. Omnik, a plastic bag conversion company in Lagos (www.omnik.biz), handles the granulation of plastics. These are then taken to a Dow testing facility in Spain for analysis. The project has employed 200 more waste pickers and is expected to raise the income of 8000 waste pickers.

PETCO South Africa

PETCO was set up as a collaboration between Coca-Cola company and various industry leaders. This company leads the recycling of PET plastics in South Africa. Its operation is partly supported by the voluntary fee it collects from plastic companies that either import or process plastics in the country. These funds are then used in buying PET bottles from waste collectors and working with local plastic recycling companies. Since its establishment in 2004, it has reportedly collected and recycled 67% of all PET bottles used in South Africa. This is a remarkable achievement compared to a global average recycling rate of between 14 to 18% and a PET bottle recycling rate in the US and EU of 28.4% and 60% in 2016 (Pietracci, 2019). PETCO has reportedly created more than 65 thousand opportunities in plastic recycling.

The PETCO model has been replicated in other countries across Africa including Kenya, Ethiopia, and Uganda. In South Africa as of 2021, all PET Coca-Cola bottles have a 25% recycled content. Coca-Cola aims to increase this to 100%. It has also committed a 38 million USD stimulus to boost plastic recycling in Southern and Eastern Africa. In 2018 Coca-Cola collected 113% more PET bottles than PET packaging than it introduced into the market (The Coca-Cola Company, 2020). As the world becomes more aware of the adverse impact of plastic waste mismanagement on the environment and humanity, increasing recycling efforts is important to the survival of companies whose products are heavily reliant on plastic packaging. This can help improve their ESG rating and the long-term survival of the company. For instance, it is in the interest of Coca-Cola that their bottles are collected and recycled as images of Coca-Cola bottles littering the environment give a negative image of the company. Engaging in recycling also gives the company direct access to the feedstock it needs for the production of its packaging; the PET bottles.

African Plastics Recycling Alliance

This was announced in March 2019 (Diageo, 2019). Its members include companies like Nestle, Unilever, Diageo, and Coca-Cola. It has the goal of supporting plastic recycling in countries in Sub-Saharan Africa towards generating jobs and investments. This it proposes to achieve through stimulating growth in the plastic recycling market, raising the demand for recycled plastics, and developing a more circular plastic economy.

Earlier in 2017, the African Circular Economy Alliance was launched by the World Economic Forum (2021a, b). This had a wider goal of promoting a circular economy in Africa that would yield economic growth and create jobs while having a positive impact on the environment. Driving economic growth that is inclusive of recycling plastics, as well as other materials, is part of the agenda of the African Circular Economy Alliance.

PRI Uganda

PRI is the leading plastic waste processor in Uganda. It reported 2,300 tonnes of plastic waste in 2018 which rose to over 5000 tonnes the following year. The plastic waste collected and recycled is processed into a range of products, which includes polyester fiber for clothing, PET bottles, and plastic sheets. The company handles collection, sorting, washing, and shredding. The recycled plastics are then either sold locally or exported to manufacturers who process them into other products.

Some Plastic Recycling Innovations Across Africa

Roof Tiles from Recycled Plastics in Kenya

Eco Blocks and Tiles Company in 2018 became the first company to manufacture roof tiles from recycled plastic and glass waste (Bhalla, 2019). The tiles show superior properties to those made from clay or concrete in terms of durability, rain resistance, ease of installation and transportation whiles having comparable costs. This application of recycled plastic keeps the plastic in use and away from the waste stream for several decades. Another added benefit is that the use of this alternative material for roofing tiles reduces the use of concrete. Despite their typically long-term use, studies have also been carried out on the end-of-life recycling of roof tiles made from recycled plastics (Balcom et al., 2021). Such studies are important for a circular plastics economy where the end-use of products is being considered before the product reaches the end of its useful life.

Floor Tiles from Recycled Tyres in Nigeria

Although this chapter has focused more on plastics recycling, rubber tires are another type PF polymers that are used and discarded on a large scale and pose environmental risks. Freetown waste management company in Nigeria is the first to set up commercial rubber tiles production from recycled car tires (WEF, 2021a, b).

Recycled Plastics Used in Roads in Morrocco

Researchers in Morocco have explored the use of recycled plastics in road construction. The study carried out as a collaboration between the Mohammed V University and the Technical Centre for Plastics and Rubber in Morocco assessed the availability of plastic waste in Morocco and the potential for use in road construction (Akkouri et al., 2020). The study then compared the properties of bitumen blended with different types of plastics with conventional bitumen in road construction. The findings suggest that the use of recycled plastics can improve the binding properties of bitumen in road construction.

Woven Recycled Plastics Floor Mat in Senegal

Colorful woven mats made of plastics are common in Senegal. The company Sosenap in Senegal recycles plastics into long strips that are then manually woven into mats and carpets (Maclean, 2022). These woven plastic mats are also seen in other parts of the continent.

Apart from the examples discussed above, various creative ways plastics and other polymers like rubber are reused and recycled can be observed across different parts of Africa. Figures 6.4 and 6.5 show the examples of ways plastic and polymers are reused and recycled as observed in Lagos and Dakar. Worn-out tires are reused as bases for mounting equipment for pumping tires and bottle caps are used in designing concrete floors.

Can Plastic Recycling be a Unique Opportunity for Africa?

The most common commodity plastics today are made from hydrocarbon-based fossil fuels. Based on current rates, it is projected that around 20% of the global oil production will go into plastics production by 2050 (UNEP, 2018). Seven African countries belong to the Organization of the Petroleum Exporting Countries whose members are currently listed as the Islamic Republic of Iran, Iraq, Kuwait, Saudi

Fig. 6.4 Waste car tires reused for mounting generators for pumping tires and also serve as a recognized symbol of the service being rendered. Lagos Nigeria. June 2022

Arabia, Venezuela, Qatar, Indonesia, Libya, The United Emirates, Algeria, Nigeria, Ecuador, Gabon, Angola, Equatorial Guinea, and Congo (OPEC, 2022). Around 6.9 million barrels of oil are produced daily within Africa (Statista, 2021) with Libya, Nigeria, Algeria, Angola, and Sudan together producing around 90% of the crude oil in the continent. Despite this, 38 countries out of 53 are net oil importers (African Development Bank, 2009). There is a need for an alternative source of feedstock for plastic production. Perhaps one that is more accessible, that does not require complex and potentially hazardous drilling. Even better if it is simply floating around the rivers. With the Nile and Niger rivers being amongst the top 10 rivers through which 90% of plastic waste flows into the world's oceans, countries around these regions can turn this into new feedstock for energy, chemicals, and plastics production.

Researchers within Africa are exploring ways to improve the yield from pyrolysis to obtain the maximum amount of oil from plastic waste gram for gram (Odejobi et al., 2020). One approach is to study how the feedstock composition of different materials and different types of plastics affects the yield from pyrolysis. For example, a 1:1 mixture of co-pyrolysis of LDPE and rice straw achieves a yield of around 61 wt% (Donaj, 2011). As the technology advances further the efficiency is likely to improve. There is also the possibility of mechanical plastic recycling processes being

Fig. 6.5 Plastic bottle caps used as designs on concrete steps at a beach in Dakar, Senegal. December 2019

powered with energy produced from the chemical recycling of plastic (Gopinath et al., 2020). Although this technology is still in its infancy, powering recycling with 100% renewable energy can bring down the emission by 51% (Zheng & Suh, 2019). Thus moving plastic recycling to become a potentially net-zero process.

While crude oil discovery, exploration, extraction, and refining require large capital investments, energy recovery from plastics can be done on a much smaller scale. This can be a unique opportunity for entrepreneurs in Africa to develop sustainable energy internally without depending on imports. This would require either individual or organizational investment in developing technical capacity in this field.

At a large scale, processes require more sophisticated controls. This doesn't imply that small-scale operations are to compromise on quality and safety standards. For example, production within a small vessel of fewer than 10 L can make use of low-cost temperature probes and pressure gauges, gaskets, and other parts that fall within a low budget while still maintaining a high level of safety. Where this is scaled up the risk becomes higher and more sophisticated control systems are required.

In plastic recycling, one of the key challenges to scaling up is actually at the earlier stages of the collection and sorting of plastics. Getting the plastics either before they get thrown out or recovering them from the sea, drainage, streets, and other locations incur cost and time. A small-scale, low-risk enterprise can be better at addressing this challenge using manual labor or developing local collection schemes. Large-scale recycling makes use of automated sorting machinery, which requires costly technical labor and in many cases, this might involve employing foreign staff and shipping in more equipment which will then require more technical labor for regular maintenance. Small-scale businesses have always been known to be a seedbed of innovation. Developing the recycling technology from the grassroots promotes better understanding and familiarization of the fundamentals of recycling technology. This gives room for homegrown technologies tailored to the need of the country to emerge. Therefore taking advantage of the relatively lower labor cost, Africa can grow a decentralized recycling industry by focusing on creating numerous small-scale businesses producing diverse products from recycled plastics rather than depending on large-scale multinationals.

Defining a New Age of Plastics in Africa

Plastics are not inherently hazardous materials to the environment. In fact, quite the contrary when we consider the life-saving applications of plastics from providing portable and affordable access to clean water to personal protective gear, and medical devices among many others. The global plastic problem lies in the misuse and mismanagement of plastic. Recycling offers the possibility of optimally harnessing the benefits of what was, and still is, a miracle material. Here, we look at Africa's positioning in the plastic recycling realm and explore the role African countries can play in redefining a new age of plastics through its recycling industry.

Already there are informal and formal recyclers across Africa converting recycled plastics into products such as roofing tiles, pavers, jewelry, blocks, and art. A search on the web and social media reveals a wide range of several informal plastic recycling businesses. Example products include drawing boards produced from recycled plastics, artworks from recycled plastics, and jewelry from recycled plastics. Many of these use small-scale production facilities aided by creative input. Below are some examples of products made from different recycled plastics based on information from across the web in Table 6.1.

The ultimate ideal of the plastic industry is a material that can be used endlessly without losing its value, and it simply goes away or becomes something equally as valuable once one is done with it. The durability of plastics combined with other properties like low density, water repellent, and low production cost compared to other materials, make plastics very attractive in endless applications. On the other hand, the durability of plastics becomes a problem when the technology or the system

Table 6.1 Examples of various products made from recycled plastics across Africa based on information from across the web from various websites and social media pages of plastic recycling businesses and products as of June 2022

Recycled product	Plastic type
Cup coaster, jewelry, drawing boards, fine art	HDPE, PP
Woven mats, bags, bins	PET bottles
High fashion outfits	PP woven matted bags commonly referred to in Nigeria as "Ghana must go"
Photoframes	Plastic bottle caps
Zip round pencil case	PET bottles
Eco brick couch	PET bottles
Lanterns	PET bottles, woven plastic bags
Hammock	Woven plastic bags
Water bottle holder	Woven plastic bags

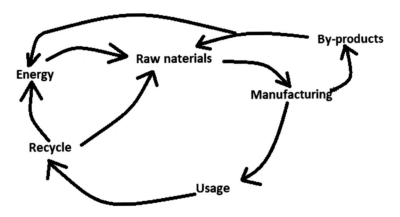

Fig. 6.6 A schematic diagram representing the ideal plastic cycle

is not put in place for the proper use and processing of post-consumer plastic and polymers. The ideal plastic cycle is one where the material is produced sustainably, used repeatedly, and, at the end of life, converted back into raw materials to repeat the cycle perpetually. This is illustrated in Fig. 6.6.

Conclusion

While hardly any country in the African continent can boast of the technological advancement in recycling that exists in, for example, France and Japan, the global plastic crisis could possibly pose a unique opportunity for Africa to lead the world in redefining the age of plastics as one of abundance, sustainability, and endless possibilities. This can be done by adopting a strategic approach to plastic recycling as discussed herein. Some small-scale industries have developed innovative approaches to plastic recycling and are having a notable impact in the region. Thus far the plastic recycling industry in the African region is predominantly made of small-scale enterprises and informal businesses. The reuse of some plastic and polymer products like PET bottles and tires has become commonplace and embedded in the commerce of some cities like Lagos. From observations and reported trends, there are indications that more plastics get recycled than the official figures reported.

References

African Development Bank. (2009). *Oil and gas in Africa* (pp. 1-2). Oxford University Press. ISBN 978-0-19-956578-8.

Akkouri, N., Baba, K., Simou, S., Nassereddin, A., & Nounah, A. (2020) The impact of recycled plastic waste in Morocco on Bitumen physical and rheological properties. In *GeoMEast* (pp 131–145). SUCI.

Balcom, P., Cabrera, J. M., & Carey, V. P. (2021). Extended energy sustainability analysis comparing the environmental impact of disposal methods for waste plastic roof tiles in Uganda. *Development Engineering., 6*, 100068.

Bhalla, N. (2019). *Kenyan scientist uses throw-away plastics to build homes.* Thomson Reuters Foundation, 13 Dec 2019

Blanco, I., Loisi, R. V., Sica, C., Schettini, E., & Vox, G. (2018). Agricultural plastic waste mapping using GIS: A case study in Italy. *Resources, Conservation and Recycling, 137*, 229–242.

Brandon, J. A., Freibott, A., & Sala, L. M. (2019). Patterns of suspended and salp-ingested microplastic debris in the North Pacific investigated with epifluorescence microscopy. *Limnologu and Oceanography Letters., 5*(1), 46–53.

Diageo. (2019). The African plastics recycling alliance- waste and opportunity. *Feature*, 26 March 2019. Retrieved February 2, 2022, from https://www.diageo.com/en/news-and-media/features/the-africa-plastics-recycling-alliance-waste-and-opportunity/

Donaj, P. J. (2011). *Conversion of biomass and waste using highly preheated agents for materials and energy recovery.* Doctoral Dissertation. Royal Institute of Technology School of Industrial Engineering and Management. ISBN 978-91-7501-033-5.

Ellen MacArthur Foundation, United Nations. (2021). The global commitment 2021: Progress report.

Fullana, F. A., & Lozano, M. A. (2015). *Method for removing ink printed on plastic films.* European Patent EP2832459A4.

Geyer, R., Jambeck, J., & Law, K. (2017). Production, use, and the fate of all plastics ever made. *Science Advances, 3*(7), e1700782. https://doi.org/10.1126/sciadv.1700782

Gopinath, K. P., Nagarajan, M. V., Krishnan, A., & Malolan, R. (2020). A critical review on the influence of energy, environmental and economic factors on various processes used to handle and recycle plastic wastes: Development of a comprehensive index. *Journal of Cleaner Production., 274*, 123031.

Grandview Research. (2021). Post-consumer recycled plastics in consumer electronics market size, share & trends analysis report by source (Non-bottle rigid, bottles) by-product (PC, PC/ABS, PET, PS, PP, ABS), by application, by region, and segment forecasts, 2021–2028. Report ID: GVR-4–68039–604–2 pp 1–140

Hoornweg, D., & Bhada-Tata, P. (2012). *What a waste: A global review of solid waste management.* World Bank. Urban Development Series Knowledge paper 15.

Jambeck, J., Hardesty, B. D., Brooks, A. L., Friend, T., Teeleki, K., Fabres, J., Beaudoin, Y., Bamba, A., Francis, J., Ribbink, A. J., Baleta, T., Bouwman, H., Knox, J., & Wilcox, C. (2018). Challenges and emerging solutions to the land-based plastic waste issue in Africa. *Marine Policy, 96*, 256–263.

Ko, S., Kwon, J. Y., Lee, U. J., & Jeon, Y. P. (2020). Preparation of synthetic graphite from waste PET plastic. *Journal of Industrial and Engineering Chemistry, 83*, 449–458.

Kyale, K., Prowe, E. F. A., Chien, C. T., Landolfi, A., & Oschiles, A. (2021). Zooplankton grazing of microplastic can accelerate global loss of ocean oxygen. *Nature Communications, 12*, 2358.

Lebreton, L., & Andrady, A. (2019). Future scenarios of global plastic waste generation and disposal. *Palgrave Communications, 5*, 6. https://doi.org/10.1057/s41599-018-0212-7

Ma, F., Liu, L., Wang, X., Jing, M., Tan, W., & Hao, X. (2020). Rapid production of few-layer graphene for energy storage via dry exfoliation of expansible graphite. *Composites Science and Technology, 185*, 107895.

Maclean, R. (2022), "Everyone's looking for plastic", as waste rises so does recycling. *The New York Times,* January 31 2022. Retrieved February 02, 2022

Milbrandt, A., Coney, K., Badgett, A., & Beckham, G. T. (2022). Quantification and evaluation of plastic waste in the United States. *Resources, Conservation and Recycling, 183*, 106363.

Odejobi, O. J., Oladunni, A. A., Sonibare, J. A., & Abegunrin, I. O. (2020). Oil yield optimization from co-pyrolysis of low-density polyethylene (LDPE), polystyrene (PS) and polyethylene terephthalate (PET) using simplex lattice mixture design. *Fuel Communications, 2–5*, 100006.

OECD. (2018). Improving plastics management: Trends, policy responses, and the role of international cooperation and trade: Policy perspectives. OECD Environment Policy Paper NO. 12. ISSN 2309-7841

OPEC Member countries, (2022). Retrieved February 11, 2022, from https://www.opec.org/opec_web/en/about_us/25.htm

Pietracci, B. (2019). *5 ways Coca-Cola is cleaning up its plastic footprint in Africa.* World Economic Forum on Africa.

Powell, J. T., Townsend, T. G., & Zimmerman, B. (2016). Estimates of solid waste disposal rates and reduction targets for landfill gas emissions. *Nature Climate Change, 6*, 162–165.

Priyanka, M., & Dey, S. (2021). Ruminal impaction due to plastic materials - an increasing threat to ruminants and its impact on human health in developing countries. *Veterinary World, 11*(9), 1307–1315.

Rehan, M., Nizami, A. S., Asam, Z. Z., Ouda, O. K. M., Gardy, J., Raza, G., Naqvi, M., & Ismail, I. M. (2017). Waste to energy: A case study of Madinah City. *Energy Procedia, 142*, 688–693.

Sadan, Z., & De Kock, L. (2020). *Plastics facts and futures: Moving beyond pollution management towards a circular plastics economy in South Africa.* WWF South Africa.

Statista. (2021). Main oil-producing countries in Africa 2020. Retrieved January 21, 2022.

The Coca-Cola Company. (2020). World without waste report.

UNEP (2018). SINGLE-USE PLASTICS: A Roadmap for Sustainability (Rev. ed., pp. vi; 6): pp 4

Wen, Z., Xie, Y., Chen, M., & Dinga, C. (2021). China's plastic import ban increases prospects of environmental impact mitigation of plastic waste trade flow worlwide. *Nature Communications, 12*, 425.

Wilson, D., et al. (2009). Building recycling rates through the informal sector. *Waste Management, 29*(2), 629–635. https://doi.org/10.1016/J.WASMAN.2008.06.016

World Economic Forum. (2021a). Five big bets for the circular economy in Africa. African circular economy alliance. Insight report April 2021a

World Economic Forum. (2021b). In Nigeria, this recycling plant turns old tyres into floor tiles.

Zheng, J., & Suh, S. (2019). Strategies to reduce the global carbon footprint of plastics. *Nature Climate Change, 9*, 374–378.

Chapter 7
Biopolymer Production and Applications

Abstract Biopolymers are widely available in the natural form and can be isolated or synthesized from fossil or nonfossil-derived raw materials. Within the African region, biopolymers like natural rubber and cotton largely get exported as raw materials. In recent years, some local and multinational companies within Africa have ventured into bioplastics production using raw materials like starch. Looking at biopolymers from a broad perspective including synthesized bioplastics such as PHA to polymers used in hair extensions and wigs gives a more inclusive picture of the impacts and potential of the biopolymer industry in Africa. The global trade routes and the supply chain mean that biopolymer products produced in other parts of the world will become available in the African region if no deliberate hindrances to importation are imposed. There are available biopolymers to meet diverse needs that are currently being met in large part by fossil-derived biopolymers.

Keywords Biopolymers · Bioplastics · PHA · Hair waste · Keratin · Silkworm

Biopolymer Production, Types, and Classifications

The term biopolymers here refer to polymers that are obtained from materials of once-living organisms. Examples are cellulose, natural rubber, starch, and proteins. Like synthetic polymers, they are large molecules made up of repeating units of covalently bonded units of smaller molecules. Some biopolymers can be mechanically processed into products in their natural form. For example, cellulose-containing jute and cotton fibers can be processed into fibers and yarns for weaving into products. Biopolymers can also be chemically modified to make them processible into other forms. For example, chemically modifying cellulose by reacting with cellulose acetate and acetic anhydride to obtain cellulose acetic can then be processed into films. While many biopolymers are biodegradable, some chemical processing can alter the biodegradability of the polymers. An example is the vulcanization of natural rubber in the presence of sulfur. The vulcanization process results in crosslinking of the polymer chains, essentially forming a network of chains that are less prone to degradation.

Bioplastics can be biodegradable or nonbiodegradable even where the raw material used in their production is bio-based. Polylactic acid, for example, is a biopolymer that is a bioplastic derived from lactic acid obtained from the fermentation of sugars, which are obtained from carbohydrates like sugarcane and corn (Groot et al., 2010). This bioplastic is bio-derived and biodegradable. Polyethylene is derived from ethylene that is synthesized from ethanol, which was obtained from the fermentation of sugarcane, however, is an example of bioplastic that is bio-derived but not biodegradable.

Biopolymers include polysaccharide, proteins, polyesters, polyisoprenes, and other polymers, which exists in various forms. Biopolymers are not new, natural biopolymers have been in existence since the beginning of life on the planet. PLA, for example, a polyester biopolymer synthesized by condensation polymerization of lactic acid, was first synthesized in 1845 by Theophile-Jules Pelouze (Benninga, 1990). To compare with fossil-derived plastics like polypropylene that was discovered in 1951 by Paul Hogan and Robert L Banks while working at the laboratory of the Philips Petroleum Company (ACS, 1999).

The next subsections discuss some biopolymers of interest and their production/isolation processes. This gives a picture of the level of technology and processes involved in the production of biopolymers using the selected examples. Subsequent sections discuss the global bioplastic market and then bioplastic production in Africa. The chapter then goes on to discuss perspectives on the biopolymer market in Africa by looking at different applications and their relevance to the region. Other biopolymers, production, and applications are discussed in the previous text (Olatunji, 2020).

PHA from Bacteria

Polyhydroxyalkanoates (PHAs) are biodegradable bio-derived polymers produced by bacteria. These bacteria aerobically ferment carbon-containing materials and produce PHAs as by-products. The industrial production of these requires isolating these bacteria and cultivating them in a system that encourages the metabolic pathways that yield optimal production of PHA. The PHAs must then be separated from the cell of the bacteria. This process requires advanced separation techniques. Currently, PHAs are the most expensive biopolymers in production costing an estimated 5–6 USD per kg. The production cost of fossil-derived plastics like polypropylene and polyethylenes is estimated between 0.5 and 2 USD per kg produced (Chanprateep, 2010).

PHAs are polyesters with properties comparable to those of the widely used commodity plastics polypropylene and polyethylene. They can be thermoplastics or elastomeric depending on the monomers that make up the polymer. Despite the advantage of being biodegradable and bio-derived, the relatively higher cost compared to these commodity plastics is prohibitive to its application as a commodity plastic. Research is ongoing into exploring more cost-effective products in order to make PHA a feasible alternative plastic for packaging since packaging forms a huge chunk of the application of commodity plastics like polypropylene and polyethylenes. Beyond packaging PHAs are applicable as bone replacements surgical pins, wound dressings, and other medical and biomedical applications.

Typically pure microbial cultures are used for the production of PHA. This requires a sterile environment to prevent contamination by other microorganisms that may hijack the system and outcompete the bacteria and produce undesired products. Bacteria such as *Escherichia coli* and *Cupriavidus necator* are used. In recent years, research studies have emerged on the use of mixed microbial culture (Ortelli et al., 2019). This does not require sterile conditions and potentially reduces the cost of production. The use of cheaper carbon sources as substrates is also being explored. However, the use of mixed culture calls for more advanced separation and purification methods, and making the process cost-effective depends on the automation of the process. For example, the use of titanium oxide and silver nanoparticles for disruption of bacteria cells for the extraction of the PHAs produced. These processes require a high level of skill in the field.

Silk Fibroin and Silk Sericin from Silkworms

Silk fibroin is a natural polymer that is produced by the *Bombyx mori* insect. It is a protein fiber that has today found a wide range of applications that are attributed to its properties such as elasticity, mechanical resistance biodegradability, and transparency. Silk fibroin can be solution-processed using conventional polymer processing techniques, and it is extracted using a water-based process that makes its manufacturing potentially sustainable.

Silk fibroin can be extracted as either native silk fibroin or regenerated silk fibroin. The former is obtained from the silk glands of the silkworm while the latter is obtained from the spun cocoons of the silk worms. Native silk is obtained by separating the glands from the silkworm through dissection just before the silkworms begin to spin their cocoons. This process, therefore, obtains the silk fibroin in its native form with minimal damage to its structure as a result of heat or chemical treatment. Regenerated silk fibroin is obtained by boiling the cocoons in 0.02 M concentration of $NaCO_3$ for 45 min. This is the degumming process that removes the alkali-soluble sericin and lipids. The degummed fibers are then rinsed in water to remove residues. This is followed by dissolving in a 9.3 M solution of LiBr at a temperature of 60 °C for a duration of 6 h. This dissolves the silk fibroins. To purify, the silk fibroins dialysis is carried out against distilled water for a period of 48 h followed by centrifugation at 14,000 rpm for 20 min (Yavuz et al., 2020). This results in an aqueous silk fibroin solution of about 6–8 w/v%. This solution is then stored for further processing. A recommended storage temperature is at 4 °C. The solution can then be further processed into desired forms such as films or fibers using polymer processing techniques such as film casting, wet spinning, or spin coating. For example, in the production of microneedles patches used for drug and vaccine delivery, studies have shown that silk fibroin microneedles can provide sustained release of active drug for up to 1 year when the silk fibroin was processed into drug-loaded microparticles (Yavuz et al., 2020).

Silk sericin is obtained in the degumming stage of silk fibroin extraction. It comprises around 25–30% of the silk protein. The sericin alongside lipids forms the continuous phase in which the silk fibroin is enveloped to form the cocoon structure. Sericin being more soluble is first removed. Further purification removes lipids and other non-silk residues. In conventional silk processing, sericin is mostly regarded as a waste. However, it does have some valuable uses. Some recent studies have explored coating synthetic fibers such as polyesters with silk sericin for improved texture (Kumar et al., 2020). Its antimicrobial properties, UV resistance, and resistance to oxidation also make it attractive in use as textiles in hygienic and medical products (Joshi et al., 2010).

Nanocellulose from Plants and Bacteria

Cellulose is a polysaccharide made up of glucose units. Cellulose is present in the cell wall of all plants and other photosynthesizing organisms such as bacteria, algae, and the ascites family of marine animals (Olatunji, 2016). Considering the abundance of plants on earth, cellulose is said to be the most abundant known polymer on earth. Application of cellulose includes clothing and textiles, pulp and paper, and food. Nanocellulose can be obtained from either the bottom-up or top-down approach. The top-down approach involves breaking down existing cellulose fibers into the nanostructure. The bottom-up approach requires building up cellulose nanostructures from carbon sources with the aid of microorganisms.

One of the dimensional attributes of nanocellulose fibers is their long aspect ratios, with diameters typically between three and four nanometers while their length is up to several microns. Cellulose is biodegradable, low density, and renewable. It is also an abundant material that makes it attractive for commercial production. Nanocellulose-based materials have demonstrated remarkable mechanical properties that make them well suited for several applications.

Production of nanocellulose can be made more cost-effective by carrying out nano-fibrillation simultaneously with compounding. Cellulose nanofibers are available commercially in sizes ranging between 5 and 100 nm. One of the raw materials for nanocellulose production is wood pulp from which nanocellulose is obtained by high-speed homogenization at a speed of 37,000 rpm for 30 min. A study suggests a wood pulp concentration of 0.7 wt% (Uetani & Yano, 2011). The nanocellulose produced in the process is in the form of an emulsion where water is the common carrier fluid. This high-speed homogenization is a top-down approach. Wood pulp can be, for example, a by-product of paper production. Such sources improve the sustainability of the process. Homogenization can be carried out with stone grinders, rotating grinders, or high-intensity ultrasounds, all of which are suited to achieve nano-fibrillated cellulose fibers.

In the buttom-up approach of nanocellulose production, bacteria fermentation of low molecular weight sugar is carried out to produce bacteria cellulose. The resulting nanocellulose from this process has superior purity. Albeit more expensive, it is important in applications such as biomedical applications like tissue engineering (Kuhnt & Camarero-Espinosa, 2021) where purity is required.

Collagen from Animal Waste

Collagen is a fibrous protein present in the connective tissue of animals. It is present in bones, skins, scales, and other tissues. In bone, for example, collagen and other proteins are present alongside the bone minerals and hydroxyapatite. Collagen composition in bone is around 90%, although the exact composition varies with age, gender, anatomic region, ethnicity, and physiological condition (Boskey, 2015). In humans, the skin comprises 78% collagen and 30% of proteins in animals are collagen (Shoulders & Raines, 2009). Collagen provides the tissues within which it is present with elasticity. It is also required for the thermal and mechanical stability of tissues. Collagen comprises three polypeptide chains that link together in distinctive ways to form the fiber structure determined by the sequence of amino acids in each chain. Different sources of collagen, therefore, tend to have different secondary structures, which is the manner in which the peptide chain interacts to form a fiber. There are up to 28 different types of collagen known, however, two main types are more prominent; type I and type II collagen. Aquatic invertebrates such as jellyfish, sea urchins, squids, and starfish have more type I collagen (Benedetto et al., 2012; Tan et al., 2013; Delphi et al., 2016). Fish scales, skins, and fins also comprise mainly this type of collagen. Type I collagen is also more abundant in human tissues. Type II collagen is present in Whale Sharks (Jeevithan et al., 2015). A common feature of collagens is the right-handed triple helix structure and the fact that every third amino acid on the polypeptide chain is glycine. Type I collagen can be distinguished from type II collagen is the type of alpha chains that make up the 3 peptide chains of the triple helix structure. The polypeptide chains that make up the collagen can be identical, built up with the same sequence of amino acids or they could be different. The former is referred to as homotrimeric while the latter is referred to as heterotrimeric collagen (Walker et al., 2021). Type I collagen comprises two alpha-1 and one alpha-2 chain (Chang et al., 2012) hence heterotrimeric. Homotrimeric collagen is more common in embryonic tissues, cancer or fibrosis tissue, or in certain genetic disorders. Type II collagen comprises three alpha-1 chains (Sharma et al., 2017).

Collagen has great potential to be a sustainable polymer material since it is often sourced from the waste product from food processing, for example, the bones from fish filleting and the skin from beef processing. While collagen might not have the turnover rate of cellulose, the most abundant polymer known, it is present in all animals both on land and at sea, therefore, it is relatively abundant in nature.

Applications of collagen include skin mimicking materials in biomedical engineering, wound healing, bone regeneration, biomedical implants, transdermal drug delivery devices, food, cosmetics, and skincare. The global collagen market has an estimated value of 8.36 billion USD as of 2020. With a cumulative annual growth rate of 9% between 2020 and 2028, the global collagen market will be worth 16.7 billion USD by the year 2028 (Grandview Research, 2021). Its relatively high growth rate can be attributed to its use in food, cosmetics, healthcare, and wellness products both being major industries. Collagen is extracted from the tissue by various means. In one method reported (Olatunji & Denloye, 2017), extraction of collagen from the scales of fish, the scales are washed with water to get rid of debris. This is then followed by washing in 0.1 M of sodium hydroxide with a liquid to solid ratio of 1:15. The scales are emersed in the alkali solution for 1 h while stirring every 15 min. This process removes lipids and other noncollagenous proteins from the surface of the scales. This is followed by draining off the liquid and repeated washing until all the alkali is removed. After this, the scales are transferred to the vessel for extraction that is typically a metal vessel made up of stainless steel. Distilled water is added in a solid to wet scales weight ratio of 1–10 weight to volume. The vessel is closed and heated to 80 °C for 8 h. The time and temperature of the process affect the yield. It has been shown that temperatures should be above 70 °C over a duration of 3 h to get a significant yield (Olatunji & Denloye, 2017). Figure 7.1 summarizes the processes involved in the extraction of hydrolyzed collagen from fish scales. The arrows show the inputs and outputs into and from the process steps.

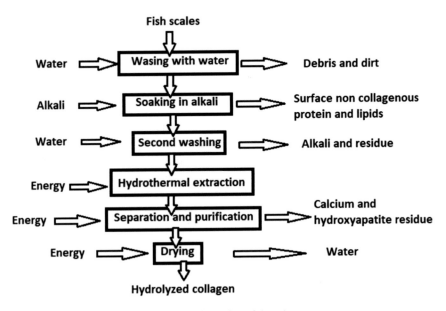

Fig. 7.1 Process chart for extraction of Collagen from fish scales

Global and Regional Bioplastics Industry

Currently, the world only has the capacity to produce around 2.42 million tonnes of bioplastics as of 2021. This is projected to rise to 7.59 million tonnes by 2026 (European Bioplastics, 2021). This figure represents less than 1% of the global plastics production (around 4 million tonnes annually). These include plastics sourced from biological sources and biodegradable plastics. Bio-based plastics that are nonbiodegradable make up 36% of bioplastics production while biodegradable plastics make up 63% of global bioplastics production (European Bioplastics, 2022). This higher production rate of biodegradable plastics could be a welcomed trend since the bio-based plastics still possess the same attributes as the fossil-based plastics and their use still comes with the same after-use challenges of their nonbiodegradability despite being bio-derived.

The cumulative annual growth rate (CAGR) of bioplastics in the Middle East and Africa is expected to reach 8.87% between 2019 and 2027 (Triton Market Research, 2019). The growth in the bioplastic market is driven by the introduction of plastic bans within the past decade with more African countries introducing bans than levies. At the current level of the technology, bioplastics prove more expensive than conventional non-biodegradable fossil-derived plastics. This is a major challenge as bioplastics need to be an affordable alternative for them to be widely adopted.

While the bioplastic product itself might be biodegradable and cause no harm to the environment. The process of production must also be carefully assessed for its environmental friendliness and sustainability. For example, burning fossil fuel to run the production plant and polluting the environment with the solvents and byproducts of the production process of bioplastics might defeat the point.

The biodegradable plastics in the market today include PLA, PCL, PBS, PHA, starch, and cellulose-based bioplastics. They are used as packaging, fibers, agriculture, or molded into products. The packaging can be in the form of films, bags, or boxes for food and non-food products. The fibers can be as fillers in composites, spun into yarns which are made into ropes or fabrics. In agriculture, they can be used as mulch, planters, tools, and other applications. Molded products can range from toys to medical implants. In recent years PLA, PBAT, PBS, and bio-based polyamides top of the chart in the global bioplastics market. Biobased polyolefins are also increasing in production. 64% of bioplastics produced globally are biodegradable while the bio-based but nonbiodegradable bioplastics make up around 36% of global bioplastics production (European Bioplastics, 2021). The bio-based non-biodegradable plastics serve as an alternative non-fossil-derived plastics, which makes them more sustainable.

Bioplastic Producers in Africa

In South Africa, Bonnie Bui Bags sells a range of plastic packaging products, which includes bags, cling wrap, straws, and cutlery. They also produce tote bags and hairdressing capes. All the products are biodegradable and biocompatible. Some will also dissolve in hot water. Although the source of energy used to heat the water and the possible emission that results should be weighed against the option of letting it compost at a lower temperature in the compost. Other companies involved in bioplastic production and supply in South Africa include The Really Great Material, Bonnie Bio Bags, Protea Chemicals, and Prime Plus Packaging are examples of bioplastic manufacturers in the country.

Bonus Industries Rwanda specializes in paper packaging as an alternative to plastic packaging. The company produces paper shopping bags, flour packaging, bread packaging, confectionery, and seed bags among others. PaperBags Ethiopia produces biodegradable stand-up pouches, gusset bags, pillow pouches, coffee bags, cold seal rolls, spout pouches, and a range of other biodegradable packaging solutions. It is located in Addis Ababa in Ethiopia and has over three decades of experience in flexible packaging production.

Other companies in the bioplastic production and supply in Africa include CareMe Bioplastics in Rwanda. Cassava, a starchy vegetable root product of the cassava plant, has gained much interest in the production of bioplastics. Cassava is a resilient and relatively easily grown plant across Africa as well as Asia and Latin America. It is a staple food in many African countries. For example, a Bali-based company presents biodegradable plastic bags produced from cassava. There have also been several research studies (Meite et al., 2021, 2022; Zoungranan et al., 2020; Oluwasina et al., 2019) on the use of starch from cassava for bioplastic production in different applications.

Some multinational companies also are key players in the African bioplastics market. For example in 2008 Natureworks LLC in the USA entered a partnership with Protea Chemicals in South Africa to distribute its Ingeo Resin across the region (NatureWorks Press Release, 2008). BASF, a multinational chemicals company has established operations in Africa for over 9 decades. It has headquarters in Midrand in South Africa, Lagos in, Nigeria, Nairobi in Kenya, and Maron in Morocco. Each represents the respective regions in the continent. Other multinational companies in the bioplastics market of Africa include Sigma Aldrich, Dow Chemicals, Mitsubishi Chemical Corporation, Tianan Biologic Material Co Ltd, and Danimer Scientific LLC among others (Triton Market Research, 2019). Some of these are companies that supply bioplastic materials for research, industrial or commercial use (Oluwasina et al., 2021).

While some bioplastic producers may not be based within Africa, some companies contacted during the research for this book were willing to supply to countries within Africa. For example, Ningbo Tianan Biologic Material Co, Ltd, a PHA (Poly(3-Hydroxybutyric acid-c-3-hydroxyvaleric acid)) producer based in China when contacted stated they were willing to supply to Nigeria. Plastic processors within Africa can therefore have access to bioplastic feedstock for the production of bioplastics-based products such as takeaway plates, packaging films, and bioplastic shopping bags. Figure 7.2 shows some biodegradable plastic plates made from bagasse on display at a Shoprite supermarket in Lagos.

Available Bioplastics and Biopolymer Resources in Africa

Top polymeric candidates to substitute plastics include cotton, jute, natural rubber, milk protein, and paper. Recent developments in the bioplastic industry are the production of plastics from waste cooking oil. Polypropylene has been successfully produced from waste cooking oil (Moretti et al., 2020). Plasticizers that are used in polyvinyl chloride have also been produced from waste cooking oil. The company Honeywell has recently announced plans to proceed to commercial production of plastics from used cooking oil and animal fat. Such new development expands the number of raw material options available for bioplastics production.

In 2019, Cote d'Ivoire was among the top three exporters of natural rubber. It exported 8.5% of the natural rubber exports for that year. Thailand and Indonesia topped the exports with 31.5% and 30%, respectively. Developing countries accounted for 94% of natural exports in 2019. As of 2018, global jute export was valued at over 210 billion USD. The natural fiber industry can be a source of employment. Take the example of Bangladesh in the Asian continent where the jute industry provides employment for around half a million people as of 2020 (WTO, 2020). The shared prospect of investing in the biopolymer industry includes an avenue to develop new businesses, create employment, and develop new skills and technologies in sustainable industries.

Other nonpolymeric materials that act as substitutes for plastics include glass, metal, and clay/ceramics. Lignin, starch, chitin, gelatin, and alginates are biomaterials that can be processed into bioplastics. Chitin is converted into chitosan that has superior film-forming properties. These films can be processed into packaging materials. Limitations of the current technology in producing packaging materials from these resources include their low water resistance compared to polyethylene. They tend to absorb moisture and become weak. They therefore might not be as versatile as the conventional plastic film. Table 7.1 lists some polymer-based materials which can serve as substitutes for fossil-derived plastics. These are either used in their natural form or processed into bioplastics. An example of a polymer resource, green seaweed is shown in Fig. 7.3a, where it is seen growing on a rock at a beach in Dakar Senegal. A closer image (Fig. 7.3b) of another side of the same rock shows some mussels attached to the rock.

Synthetic non-biodegradable plastics and polymers may well be versatile and cheap; however, considering the cost to the environment and the range of alternative materials available, it is worth considering compromising the convenience of plastics for more sustainable options. Therefore rather than using the same polypropylene for food bowls, shopping bags, cooking utensils, and ropes, the feedstock can be

(a)

(b)

Fig. 7.2 a Plant-based plastic plates found on the shelf at Shoprite Ikeja, **b** close-up image. March 2022

Table 7.1 Substitute materials for plastics and their production rate in 2018 (UN data, FAOSTAT) and some example applications

Material	Production rate in Africa (tonnes)	Application	Fossil-derived commodity plastics commonly used for same applications
Jute	8, 472 tonnes	Woven sacks for packaging cement, fertilizers, and agricultural produce	PP
Cotton Seed	4,491,964 (2016)	Fabrics	Polyester, Nylon
Starch	Angola 583,630 (2016)	Packaging films	HDPE, LDPE
(corn)	Kenya 609,700 (2016)		
Cassava	169,673,737		
Natural rubber	816,985	Shoes, mats	Polyisoprene
Wood pulp (For p)pulp & paper, corrugated paper, and cardboard	2,329,000	Packaging bags and boxes	HDPE, PS
Chitin Shrimps nonfrozen (exports from Gambia, Senegal, Kenya, and Botswana	12,089	Packaging films	LDPE, PP, HDPE, PET
Animals slaughtered	43,882,171 (heads)	Packaging films	LDPE, PP, HDPE, PET
Hide (Fresh Cattle)	982,108		
(Bones and skin for gelatin production)			
Seaweed for alginate production)		Packaging films	HDPE, LDPE
Angola	31.461		
Botswana	481.147		
Ghana	97.649		
Kenya	1.090		
Rice Husk		Cutlery and plates	PP
Rice	36,637,152 (2013)		
Coconut (Coconut fiber source)	2,042,542	Sponge	PP, Nylon
Milk protein (Casein)	46,653,629	Cushions, jewelry	PS

(a)

(b)

Fig. 7.3 a Green seaweed growing on a rock by the beach in Dakar, Senegal **b** Same rock with mussels attached to the surface. December 2019

diversified too, for example, using cotton for shopping bags, jute for ropes, and PLA for food bowls. Having a more diverse range of feedstock also allows for a more diverse market with opportunities for a wide range of skills and raw materials rather than having one company for example processing the same HDPE into 10 different products.

Lignin, for example, can be processed into a thermoplastic material that has the processability of plastic combined with the biodegradability and physical properties and appearance of wood (Nagele, 2002). Such material can be processed using conventional polymer processing techniques such as injection molding and extrusion, into packaging materials. An estimated 50 million tonnes of lignin is generated as a by-product of the pulping process. The paper and pulp industry is, therefore, a source of lignin.

Impact of the Growing Bioplastic Industry on Africa

In a December 1st press release in 2021 in Berlin, the European Bioplastics presented a growth outlook on the bioplastic industry. The industry is expected to grow three times its current rate by 2026 from 2.4 million tonnes to 7.5 million tonnes between 2021 and 2026. This can be attributed to the increasing visible impact of plastic pollution on the environment and the demand for more environmentally friendly alternatives to fossil-derived non-biodegradable plastics.

Considering that 7 African countries are among the 13 member countries of the Organisation of the Petroleum Exporting Countries (OPEC) and around 12% of global petroleum production goes towards manufacturing plastics. The switch to bioplastics is expected to affect the demand for crude oil. Much of the bioplastics are sourced from agricultural raw materials such as corn and biomass. To have a tangible means of trade in the bioplastic industry, these countries need to develop the agriculture sector to meet the demand for raw materials in the growing bioplastics industry.

Thus far the European Bioplastics suggests that the rise in production of bioplastics is not in competition with food and feed for land since only about 0.01% of the global agricultural land area is being used for cultivating raw materials for bioplastics production. Even with the projected rise, this is expected to only increase to 0.06% within the next 5 years. Nonetheless, contrary views exist that every acre of land and the resources used for growing crops for bioplastic raw materials could have been put into use for food production.

Asia remains a leading region in the global production of bioplastics. The continent produces around 50% of the global bioplastics with Europe coming second at around 25%. It is predicted that by 2026 Asia will the producing 70% of the world's bioplastics. With increasing trade between Asian and African countries in recent years, Africa is likely to be impacted by this growing bioplastic industry in Asia. This can be for example trade agreements that lead to an exchange of technical capacity in bioplastic production, bioplastic import/export agreements, or bioplastic raw materials trade.

As the bioplastic industries grow and more of the plastics circulating around the world and reaching Africa through global trade routes increasingly comprise bioplastics, this will potentially reduce the strain on the waste management system. Low technology methods such as landfilling and burying are the most readily available plastic disposal methods. These methods have associated adverse impacts such as soil contamination. The use of nontoxic biodegradable plastics and polymers will make methods such as landfilling and burying more sustainable as the plastics will degrade faster and give more room for more plastics.

Biopolymers in Diverse Applications

Polymers are everywhere. Considering that in every industry polymers are used in one form or the other and that polymers such as cellulose in plant cell walls, DNA chains, keratin in hair and collagen in muscles and so much more form the body of humans and other organisms, that is not at all an exaggerated statement. With the challenges of the current non-biodegradable polymers, biopolymers potentially serve as a pathway to a more sustainable polymer industry. The biopolymer industry is still in its early development stage. There are several opportunities in the biopolymer industry that extends to other industry sectors such as food, cosmetics, pharmaceutical, and others. With each region of the world having its own peculiarities in various sectors, here we focus on the African continent. This section reviews the established and potential application of biopolymers in the industry and how this applies to specific aspects and current issues in the region.

Food

Naturally occurring polymers like starch, cellulose, and proteins play an important role as food and energy sources. In the modern food industry where food products travel up to thousands of miles to get to the consumer, polymer processing techniques are applied to food to process and preserve them. Drying of fruits, extrusion of noodles, molding of chocolates, fermentation of rice, and other food processes are widely applied to modify the polymer components of foods to achieve taste, texture, shelf life, and other properties.

The food industry also serves as a source of bioproducts that are polymers of high value. For example, the scales of fish and the shells of crustaceans and snails are a source of collagen and chitin. These are polymers from the food industries, which get used in food and other applications. Figure 7.4, for example, shows an image of shells of snails being separated at a market in Lagos. These shells according to discussion with traders are typically discarded.

Fig. 7.4 Snail trader separating the shells from the snails at a market in Lagos. April 2022

In its recent report on the state of global food security (FAO, 2021), the FAO high-lights the need for affordable healthy diets. As of 2019, an estimated 3 billion people globally cannot afford a healthy diet. The number of people who cannot access a healthy diet in Africa increased between 2017 and 2019. The reasons include changes in weather patterns, droughts, floods, conflicts, COVID-19, and disruptions to global supply chains. Therefore, advancing the applications of biopolymers in food processing and packaging can contribute towards the goal of solving global food security and ending hunger and malnutrition by 2030. One of the areas of need that were identified for Africa was affordable nutritious food availability. Biopoly-mers are key materials that can be used to process food in order to preserve the nutrients and even fortify foods with nutrients. Examples are biopolymers used as food stabilizers, protective films, and thickeners. As discussed in other chapters of this book, biopolymers such as starch and gum arabic are produced within the continent in significant quantities. The advancement of the biopolymer industries in Africa will aid technological advancement in food processing that could in turn boost food security.

Understanding the structure and properties of polymers in foods can also advance technologies in the optimal extraction of nutrients from byproducts of food processing. For example, extraction of essential nutrients from fish processing waste and residues from fruit juice processing. For example, in the production of tomato paste, an understanding of the role of pectin and the pectinase enzyme as well as the temperature-viscosity relationship in polymer processing is essential to optimal tomato paste production (Thomson & Castilho, 2010). The tomato paste production reduces the loss of tomatoes due to spoilage. The prolonged shelf life and packaging also allow the food to be available for longer and transported further. In a study of 408 producers and consumers across Ethiopia, it was found that around 28% of tomatoes lost were at the producers, 8.6% was at the wholesaler, 2.9% was at the retailer while 7.3% was at the point of consumption such as hotels and cafes (Abera et al., 2020). These post-harvest losses were due to various factors such as handling, inappropriate storage temperature, and damage in the process of transportation. Processing such goods into more durable forms such as purees and pastes in sachets and cans can reduce food loss. Adequate processing can also ensure that the maximum amount of nutrients and taste is retained in the processing.

Drug and Vaccine Delivery

Due to their structure, many polymers are well suited to serve as drug and vaccine delivery materials. Polymers are used in the production of tablets, gel capsules, bolus, and controlled and sustained release in transdermal drug delivery patches, implants, and other drug delivery modalities. They are used for the delivery of hydrophilic and hydrophobic compounds and through various delivery routes such as oral, buccal, nasal, and transdermal among others. These polymers include synthetic and natural, biodegradable, and nonbiodegradable. Applications range from starch used in tablets, to backing films that can be made of polymers like polypropylene, and polymeric adhesives to a drug-loaded matrix made of polymers such as polyvinyl alcohol. Recent advances include pH or temperature or pressure responsive polymers, hydrogels, and novel polymerization techniques for achieving more sophisticated drug delivery goals (Liechy et al., 2010).

Among recent developments in the field of drug and vaccine delivery are microneedles for drug and vaccine delivery. Microneedles are micron-sized projections that are designed to painlessly pierce through the uppermost, nerve-free layer of the epidermis, thereby creating micro conduits through which moieties can more easily permeate the skin bypassing the impermeable top layer. Microneedle fabrication using polymers offers the benefit of mass production, and low-temperature processing that ensures that the active ingredient is not damaged in the process of production, and the use of biodegradable polymers achieve microneedles that can safely dissolve in the skin for controlled release and minimal waste generation.

Vaccines stored in the form of microneedles further have the advantage of reaching rural areas without the need for highly skilled medical expertise to administer and the fact that vaccines in dry form are potentially more resistant to temperature fluctuations in storage. Polymers such as silk fibroin, hydrolyzed collagen, starch, and many others have been used in the production of microneedles for a broad range of applications. Production techniques range from conventional polymer processing techniques such as thermal drawing to more recent techniques such as 3D printing. Microneedles have proven promising in areas of diabetes management, vaccine delivery, and pain management among others.

Another example of interest is the use of shells of oil seeds such as Argan oil seed shells. Argan shells containing polymers such as proteins and polyphenols are being explored for application in drug adsorption (Babas et al., 2022). The oil already is highly valued for cosmetic applications, finding applications in drug absorption procedures will further add to the value of the argan plant and also make the production of argan oil more sustainable by utilizing the shells.

In 2022, WHO announced Nigeria, Egypt, Kenya, Senegal, South Africa, and Tunisia as the six African countries that will be provided with tools, training, and technical capacity for the production of mRNA vaccines. This is in the move to promote the production of vaccines in the region through global technology transfer. This will move vaccine manufacturing from primarily high-income countries to low and middle-income countries, which includes many countries in Africa. It was estimated that 900 million vaccine doses are required to attain 40% vaccination of the African population; however, only 492 million doses had been secured by January 2022 and only 64% of the procured vaccines due to expiration and other reasons such as availability of syringes for administering the vaccines and mobilization of the professionals to administer them (The Lancet Infectious Diseases, 2022). Nucleic acids, recombinant proteins, and peptide sequences are polymeric materials used in vaccine production. The adjuvants used to stimulate immune response and boost the efficacy of the vaccines are also made of polymeric materials (Bose et al., 2019). Therefore, developing technical capacity and technologies in Africa should also include the application of polymers in these fields.

Wound Healing and Tissue Repair

Tissue engineering is important in improving the quality of life. Healthy and able-bodied individuals are also better able to take care of themself and contribute to society. Tissue engineering in areas such as bone replacement is important in restoring functionality to body tissues after damage. Biopolymers have found vast applications in this area. For example, the use of biopolymers as tissue scaffolds, skin grafts, intraocular lens replacement, and ear reconstruction among others. Polymers have replaced materials like metals for example hip replacement.

Properly caring for wounds resulting from cuts and other damages is important in preventing them from deteriorating and leading to infections and amputations or even fatalities. Already medicinal plants native or common to Africa have been used in wound healing across Africa and other parts of the world for centuries. Polymers such as chitosan, collagen, and hyaluronan incorporating enhanced wound healing agents such as silver nanoparticles have been shown to improve the effectiveness, applicability, and practicality of these medicinal plants and some of such products have been approved by FDA (Tyavambiza et al., 2021). By designing more effective polymer-based wound healing materials that are affordable and effective, the polymer industry can empower hospitals and healthcare centers in providing high-quality healthcare and improve patients' quality of life.

Average government spending on healthcare in sub-Saharan countries (excluding South Africa where healthcare spending is similar to that of the UK and US) is estimated to be around 90 USD per person annually. In higher-income countries, this is around 5,200 USD. Considering that the cost of for example hip and knee replacements is around 7,000–9,000 USD in the UK and around 16,000–60,000 in the US (Davies et al., 2019), further efforts are required to reduce the cost of tissue replacement. Polymers offer the advantage of affordability, personalized designs, realistic feel, better aesthetics, improved tissue attachment, and ease of fabrication. With new technologies such as 3D printing of tissue scaffolds made of polymeric materials and the use of biodegradable and biocompatible polymers, wound healing and tissue repair and regeneration procedures can be made better. Table 7.2 gives some example applications and polymers used for these wound healing and tissue repair applications.

Table 7.2 Polymers used in various tissue engineering applications

Application	Example polymers used	References
Intraocular lens replacement	Crosslinked polyisobutylene	Pinchuk (2022)
Implant for glaucoma treatment	Polystyrene block—isobutylene block—styrene	Pinchuk (2022)
Skin grafting	Polygalactic acid, polyglycolic acid, alginate, pectin, chitosan, collagen, pullulan, gelatine	Augustine et al. (2014)
Bone regeneration	Chitosan, alginates, gelatine, PDA, PLGA, PLA, PCL, PEG, PVA, PEEK	Wang et al. (2022)
Auricle regeneration	Gelatine methacrylate, polylactic acid	Tang et al. (2021)

Cosmetics

The global cosmetics market is estimated to have generated sales worth 220 billion Euros in 2019. The sales grew by 5.5% between 2018 and 2019. L'Oreal, for example, reported a consolidated sale of 29.873 billion Euros in 2019 (L'Oreal, 2019). 2021 saw a rise in global cosmetics revenue of 8 billion USD reaching 80.74 billion USD in the year. The revenue for the cosmetics industry is projected to reach 131 billion USD by 2026. This is revenue from decorative cosmetics such as lipsticks, mascaras, and face powders excluding personal hygiene and skin care products (Statista, 2022). Table 7.3 lists of L'Oreal sales% by region.

According to the 2019 L'oreal annual report, Africa and the Middle East account for 2.3% of L'oreal product sales and 2% of the global luxury cosmetics market. L'oreal, Unilever, Estee Lauder, Procter & Gamble, Shiseido, and Coty made 31.8, 22.4, 14.2, 13.2, 9.7, and 9.1 billion USD in global cosmetics sales in 2019 (L'Oreal, 2019). Table 7.3 summarizes the reported sales of L'Oreal globally, expressed in percentages by region in 2019.

L'Oreal lists the increasing scarcity of natural resources as one of the challenges facing Sub-Saharan Africa's cosmetics market (L'Oreal, 2019). According to UN Data, Morocco outputs cosmetics products worth 4,807,510,000 Dirhams (1,308,865,798 USD approximately) in 2011, that of Eritrea was 79,544,909 Nakfa. To compare, Russia's output was estimated as 129,072,486,964 Roubles and that of Australia was 3,482,000,000 Dollars. Unilever Nigeria owns cosmetics and personal care brands such as Lux, Vaseline, Pears, Close-up, Pepsodent, and Lifebuoy reported revenue of NGN 70,523,695,000 in 2021. A significant rise from the NGN 52,211,267,000 was reported in 2020 (Unilever, 2021). These sale figures include other categories such as food seasoning and laundry products. Some of these brands have been in existence for over 150 years.

Collagen, keratin, and peptides are among the ingredients listed in high-end cosmetics products sold at beauty stores in African cities like Lagos. There is increased urbanization, and this urban population is increasingly adopting the popular beauty trends promoted by social media, the entertainment, and the fashion industry in an increasingly globalized world.

Argan oil is one of the most priced oils for cosmetic applications. Crude extracts from the shells contain proteins and polyphenols that have been studied for their

Table 7.3 L'Oreal sales in 2019 by region

Region	Sales%
Western Europe	27.7
North America	25.3
The Asia Pacific	32.3
Eastern Europe	6.4
Latin America	6
Africa, Middle East	2.3

emulsifying properties (Bouhoute et al., 2020). Emulsifiers are essential ingredients for producing lotions for products such as skin moisturizers and liquid preparations like foundations. Such understanding of the polymers present in the by-products of valuable products such as argan oil can further create value and increase the output from the plant. Combustion of the shell also has potential for use as a clean fuel in, for example, incineration (Rahib et al., 2021) or chemical recycling.

Hair Care, Hair Extensions, and Hair Waste Utilization

Hair is a natural fiber that is produced by animals protruding from the surface of the skin. The main polymer in hair is Keratin a fibrous protein. The structure of the hair fiber of an animal depends on the order of the amino acids present in the protein. This is determined by the genetic instruction in the DNA of the individual. As of 2017, wigs and extensions made up around 18% of global revenue in the beauty industry. The industry is expected to grow at a 13% cumulative annual growth rate and above between 2021 and 2026. The largest consumers of wigs and hair extensions are said to be people of African descent and native Africans. The rise in income is expected to result in increased demand for wigs and hair extensions in addition to the urban image of wigs and hair extensions in the fashion and entertainment industry promoting the use of wigs and extensions and the increased reports of hair loss rate among men and women (ReportLinker, 2021). With increased urbanization in the African region, more people are likely to adopt the urban images presented by an urban culture, which presently includes makeovers and sophisticated looks with wigs and hair extensions. Therefore, the growth of the hair extensions and wig industry can be linked to other social factors and trends. Figure 7.5 shows an image of the hair extension section at a supermarket in Lagos Nigeria taken on May 13, 2022.

Recent innovations include natural fiber hair extension where for example sodium alginate and protein from Antarctic Krill is wet sun into hair fibers (Yang et al., 2017). The process of production can be made environmentally friendly without the use of toxic solvents. The resulting wet spun hair is said to sufficiently mimic human hair. Addressing rare issues such as rapunzel syndrome and synthetic hair extension causing pollution by making them biodegradable and biocompatible and nontoxic. In a case like accidental ingestion or rapunzel syndrome, the hair will degrade in the digestive system rather than accumulate and require surgery to be removed.

Other studies are also exploring using hair waste for example concrete for construction (Mangunatha et al., 2021). This study showed that the addition of waste hair fibers gave concrete improved properties. Keratin from waste human hair is also being explored for use as scaffolds in tissue engineering. When tested in cell cultures and in vivo experiments on animals, scaffolds produced from extracted keratin from human hair attained good porosity and showed good cell proliferation and tissue vascularization. The scaffold also biodegraded over time giving way to newly formed tissues (Xu et al., 2013). However, keratin extraction requires the use

Fig. 7.5 Image of hair extension on display at a section of a supermarket in Lagos, Nigeria on May 2021

of chemicals like HCl and chloroform, therefore, the environmental impact of the extraction process should be carefully weighed against the benefits.

Hair waste can also be used in the production of fertilizers and other biochemicals in a sustainable way through bacterial biodegradation (Bach et al., 2015). Through understanding the optimal conditions for the production of the enzyme keratinase by these keratin degrading bacteria, bioreactors for large-scale production of fertilizers and biochemicals from hair waste from humans and other animals offer a promising avenue to meet the demand for fertilizers and other products of hair degradation. Fertilizer production is also linked to food production according to UN data, mainland

China imported around 1.46 million tonnes of NPK fertilizer and exported just over 365 thousand tonnes. Taking Egypt as an example within Africa, Egypt imported just over 19 thousand tonnes of the same fertilizer and exported just over 5 thousand tonnes. These lower values could be linked to factors such as differences in the scale of agriculture. Hair waste can, therefore, be diverted to fertilizer production to boost domestic production of fertilizers.

Understanding hair as a polymeric material can be beneficial in several ways. Within the African concept, an understanding of the polymeric structure of African hair is important in addressing rising issues of conditions such as alopecia and meeting the demand for hair care products that adequately address the needs of hair of people of African origin. For example, one form of modern hair care that has been adopted in African hair care is the use of chemical relaxers. This process involves the application of concentrated sodium hydroxide (check the concentration of sodium hydroxide in relaxers and other ingredients), to alter the secondary structure of the hair fiber.

The interaction between the functional groups on the polymer chains that make up the hair fiber that comprises keratin protein polymers bonded by strong hydrogen bonds results in the tertiary structure of the hair, which is the curved structure. The level of curvature or curliness of the hair varies from tight to loose curls and is determined by the sequence of the amino acids, which make up the protein in the hair fiber. Alkali treatment weakens amino linkages. Therefore over time and repeated application of sodium hydroxide to the hair, it is prone to weakening. For example, as discussed earlier in the chapter, the protein in the scales of fish, collagen is reduced to gelatine, which is hydrolyzed collagen, hence shorter chain collagens. This is expected to have a similar impact on the protein keratin in hair.

An alternative is the heat treatment of hair in order to achieve a straighter appearance. This involves applying heat typically with a hair straightening device that heats the hair under pressure to a temperature between 150 and 200 °C. The heat disrupts the bending of the polymer chain caused by the interaction between the functional groups and hence straightens the hair temporarily. The effect is temporary because the new bonds that resulted in the new chain conformation are weaker than the original bonds that created the curly hair fiber structure. Therefore over time, the hair returns to its curly state. This is also accelerated by humidity.

A thorough understanding of the polymer chemistry of the different African hair types is important to develop more sustainable and better-suited hair care products for hair of African origin. Beyond this, understanding the polymer science of hair can promote a positive appreciation of African hair in its true form and prevent the use of unsustainable hair care regimes that result in increased energy and toxic chemicals consumption. For example, daily hair straightening and application of chemicals and use of nonbiodegradable hair extensions. Examples of recent studies in this area include studies on the response of hair to fatigue over several loading cycles (Ngoepe et al., 2021).

Biopolymers and Food Security

Developing the biopolymer technologies for application in the diverse industries in the African region should consider an industry that includes everyone working in the industry. This includes indigenous people, farmers, factory workers, entrepreneurs, small-scale producers, and more. The industry should consider the applications of biopolymers in production, storage, transportation, packaging, and marketing and applications of biopolymers in providing goods and services in diverse sectors. For example, it has been established that food security goes well beyond increasing production yield or providing silos for storing grains. Rather, food security requires more advanced solutions such as processing technologies that prevent post-harvest losses (Abera et al., 2020).

There are growing concerns over the impact of plastic ingestion on livestock and farm animals. When these animals ingest plastics, it accumulates in the rumen, and this eventually leads to health defects such as ruminal impaction, indigestion, and recurrent tympany among others (Priyanka & Dey, 2021). Animals that are allowed to openly graze are particularly susceptible to ingesting plastic litter. This is more common in urban areas in Africa and other developing countries where plastic littering is more common. Figure 7.6 shows an example of animals openly grazing in the town of Simawa in Nigeria. The image shows a plastic litter around the cows as they graze. These cows being unable to separate the plastics from the grass are likely to ingest some of the plastics alongside the grass and other edible matter in the litter. The dangers of plastic bag ingestion can be mitigated by replacing the nonbiodegradable plastic bags with biodegradable ones. This has an extended impact on both the livestock and the consumers of the meat and milk products of the livestock.

Conclusion

The African continent produces a variety of raw materials that can serve as substitutes for fossil-derived synthetic non-biodegradable plastics. Raw materials for bio-based synthetic biodegradable and not biodegradable plastics can also be produced from these raw materials. Many of the bioplastics producers within the content found in this study were based in South Africa with a few in Rwanda and Ethiopia. Multinational companies and other bioplastics producers have the capacity to supply bioplastics raw materials or finished products across the world. Naturally occurring polymers such as jute and cotton can replace problematic plastics such as plastic shopping bags while other polymeric raw materials such as starch and gelatin can be processed into bioplastics for applications such as food packaging films. Making a shift to biopolymer-based products replacing commodity synthetic plastics can lead to a more diversified plastic industry where multiple feedstocks are used to achieve a variety of products compared to one where the same plastic is used to produce

Fig. 7.6 Cows openly grazing in the semi-urban area of Simawa, Ogun State Nigeria may come across plastic waste like the one indicated in the black circle in the image. April 2022

various types of products. This can potentially give room for more players in the plastic industry of Africa in addition to a more sustainable plastic economy.

References

Abera, G., Ibrahim, A. M., Forsido, F. S., & Kuyu, C. G. (2020). Assessment on post-harvest losses of tomato (Lycopersicon esculentem Mill.) in selected districts of East Shewa Zone of Ethiopia using a commodity system analysis methodology. *Heliyon, 6*(4), e03749.

ACS American Chemical Society National Historic Chemical Landmarks. (1999). *Polypropylene and high-density polyethylene.* Retrieved 16 June, 2022, from http://www.acs.org/content/acs/en/education/whatischemistry/landmarks/polypropylene.html

Augustine, R., Kalarikkal, N., & Thomas, S. (2014). Advancement of wound care from grafts to bioengineered smart skin substitutes. *Progress in Biomaterials, 3*, 103–113.

Babas, H., Khachani, M., WaraI, A. S., Guessous, A., Guenbour, A., Safi, Z., Berisha, A., Bellaouchou, A., Abdelkader, Z., & Kaichouh, G. (2022). Sofosbuvir adsorption onto activated carbon derived from argan shell residue: Optimization, kinetic, thermodynamic and theoretical approaches. *Journal of Molecular Liquids, 356*, 119019.

Bach, E., Lopes, F. C., & Brandelli, A. (2015). Biodegradation of alpha and beta keratins by gram-negative bacteria. *International Biodeterioration and Biodegradation, 104*, 136–141.

Benedetto, C. D., Barbaglio, A., & Martinelo, T. (2012). Production, characterization and biocompatibility of marine collagen matrices from an alternative and sustainable source: The sea orchin Paracentrotus lividus. *Marine Drugs, 12*, 4912–4933.

Bose, R. J. C., Kim, M., & Park, H. (2019). Biodegradable polymers for modern vaccine development. *Journal of Industrial and Engineering Chemistry, 77*, 12–24.

Boskey, A. L. (2015). Bone composition: Relationship to bone fragility and antiosteoporotic drug effects. *Bonekey Reports, 4*, 710.

Bouhoute, M., Taarji, N., Vodo, S., Kobayashi, I., Zahar, M., Isoda, H., Nakajima, M., & Neves, M. A. (2020). Formation and stability of emulsions using crude extracts as natural emulsifiers from Argan shells. *Colloids and SurfacesA: Physicochemical and Engineering Aspects, 591*, 124536.

Chang, S., Shefelbine, S. I., & Buehler, M. J. (2012). Structural and mechanical differences between collagen momo- and heterotrimers; relevance for the molecular origin of brittle bone disease. *Biophysics Journal, 102*(30), 640–648.

Chanprateep, S. (2010). Current trends in biodegradable polyhydroxyalkanoates. *Journal of Bioscience and Bioengineering, 110*(6), 621–632.

Davies, P. S. E., Graham, S. M., & Harrison, J. W. (2019). Total joint replacement in Subsaharan Africa: A systematic review. *Tropical Doctor, 49*(2), 120–128.

Delphi, L., Sepehri, H., & Motevaseli, E. (2016). Collagen extracted from Persian golf squid exhibits anti-cytotoxic properties on apple pectic treated cells: Assessment in an in-vitro bioassay model. *Iranian Journal of Public Health, 45*, 1054–1063.

European Bioplastics. (2021). Press release: Global bioplastics production will more than triple within the next five years. Berlin 1 December 2021. In *16th EUBP Conference.*

European Bioplastics. (2022). *Bioplastics market data.* Retrieved 28 March, 2022, from www.european-bioplastics.org/market

FAO, IFAD, UNICEF, WFP, & WHO. (2021). In a brief to the state of food security and nutrition in the world 2021. In *Transforming food systems for food security, improved nutrition, and affordable healthy diets for all.* FAO. https://doi.org/10.4060/cb5409en

Grandview Research. (2021). *Collagen market size, share & trends analysis report by source, by product (gelatin, hydrolyzed, native, synthetic) by application (food & beverages, health-care, cosmetics), by region, and segment forecasts, 2021–2028* (Report ID: GVR-1-68038-835-0) (pp. 1–128).

Groot, W., van Krieken, J., Sliekersl, O., & De Vos, S. (2010). Production and purification of lactic acid and lactide. In R. Aura, L. Lim, S. E. Selke, & H. Tsuji (Eds.), *Poly(lactic acid) synthesis, structures, properties, processing, and applications* (pp. 3–4). Wiley.

Jeevithan, E., Zhang, J. Y., Wang, N. P., He, L., Bao, B., & Wu, W. H. (2015). Physico-chemical, antioxidant and intestinal absorption properties of whale shark type-II collagen based on its solubility with acid and pepsin. *Process Biochemistry, 50*, 463–472.

Joshi, M., Punwar, R., & Ali, W. (2010). Antimicrobial textiles for health and hygiene applications based on ecofriendly natural products. *Medical and Healthcare Textiles* (Woodhead Publishing Series in Textiles), 84–92.

Kuhnt, T., & Camarero-Espinosa, S. (2021). Additive manufacturing of nanocellulose based scaffolds for tissue engineering: Beyond a reinforcement filler. *Carbohydrate Polymers, 252*, 117159.

Kumar, M., Janani, G., Fontaine, M. J., Kaplan, D. L., & Mandal, B. B. (2020). Silk-based encapsu-lation materials to enhance pancreatic cell functions. In Ormando, et al. (Eds.), *Transplantation, bioengineering, and regeneration of the endocrine pancreas* (Vol. 2, pp. 329–337). Academic Press. ISBN: 9780128148310.

L'Oreal. (2019). *2019 annual report.*

Liechy, W. B., Kryscio, D. R., & Peppas, N. A. (2010). Polymers for drug delivery systems. *Annual Review of Chemical and Biomolecular Engineering, 1*, 149–173.

Mangunatha, M., Kvgd, B., Vengala, J., Manjunatha, L. R., Shankara, K., & Patnaikuni, C. K. (2021). Experimental study on the use of human hair as fiber to enhance the performance of concrete: A novel use to reduce the disposal challenges. *Materials Today: Proceedings, 47*(13), 3966–3972.

Meite, N., Konan, L. K., Tognonvi, M. T., Doubi, G. H. I., Gomina, M., & Oyetola, S. (2021). Properties of hydric and biodegradability of cassava starch-based bioplastics reinforced with thermally modified kaolin. *Carbohydrate Polymers, 254*, 117322.

Meite, N., Konan, L. K., Tognonvi, M. T., & Oyetola, S. (2022). Effect of metakaolin content on mechanical and water barrier properties of cassava starch films. *South African Journal of Chemical Engineering, 40*, 186–194.

Moretti, C., Junginger, M., & Shen, L. (2020). Environmental life cycle assessment of polypropylene made from used cooking oil. *Resources Conservation and Recycling, 157,* 104750.

Nagele, H. (2002). Albaform®—A thermoplastic, processable material from lignin and natural fibers. In T. Q. Hu (Ed.), *Chemical modifications, properties and usage of lignin.* Springer. 978-1-4613-5173-3.

Natureworks Press Release. (2008, October 14). *NatureWorks expands Ingeo resin distribution.* Retrieved from 10 February 2022.

Ngoepe, N. M., Cloete, E., van de Berg, C., & Khumalo, N. P. (2021). The evolving mechanical response of curly hair fibers subject to fatigue testing. *Journal of Mechanical Behaviour of Biomedical Materials, 118,* 104394.

Olatunji, O. (2020). *Aquatic biopolymers: Understanding their industrial significance and environmental implication.* Springer Series on Polymer and Composite Material. Springer Nature Switzerland. ISBN 978-3-030-34708-6.

Olatunji, O. (2016). *Natural polymers: Industry techniques and applications.* Springer. ISBN: 978-3-319-26412-7.

Olatunji, O., & Denloye, A. (2017). Temperature-dependent extraction kinetics of hydrolyzed collagen from scales of croaker fish using thermal extraction. *Food Science and Nutrition, 5,* 1015–1020.

Oluwasina, O., Akinyele, B., Olusegun, S., Oluwasina, O., & Mohallem, N. (2021). Evaluation of the effects of additives on the properties of starch-based bioplastic film. *Springer Nature Applied Sciences, 3,* 421.

Oluwasina, O. O., Olaleye, F. K., Olusegun, S. J., Oluwasina, O. O., & Mohallem, D. S. N. (2019). Influence of oxidized starch on physicomechanical, thermal properties, and atomic force micrographs of cassava starch bioplastic film. *International Journal of Biological Macromolecules, 135,* 282–293.

Ortelli, S., Costa, A. L., Torri, C., Samori, C., Galletti, P., Vineis, C., Varesano, A., Bonura, L., & Bianchi, G. (2019). Innovative and sustainable production of biopolymers. In Tolio, et al. (Eds.), *Factories of the future* (pp. 131–148). Springer. ISBN 978-3-319-94358-9.

Pinchuk, L. (2022). The use of polyisobutylene-based polymers in ophthalmology. *Bioactive Materials, 10,* 185–194.

Priyanka, M., & Dey, S. (2021). Ruminal impaction due to plastic materials—An increasing threat to ruminants and its impact on human health in developing countries. *Veterinary World, 11*(9), 1307–1315.

Rahib, Y., Boushaki, T., Sarh, B., & Chaoufi, J. (2021). Combustion and pollutant emission characteristics of argan nut shell (ANS) biomass. *Fue; Processing Technology., 213,* 106665.

ReportLinker. (2021). *Hair wigs and extensions market—Global outlook and forecast 2021–2026* (ID: 5822878).

Research and Markets. (2019). *Collagen market by product type—Global forecast to 2023* (Report ID: 4756592).

Sharma, U., Carrique, L., Le Goff, V., Mariano, N., Georges, N. R., Delome, F., Koivunen, P., Myllyharju, J., Moali, C., Aghajari, N., & Hulmes, D. J. S. (2017). Structural basis of homo and heterodimerization of collagen I. *Nature Communications, 8,* 14671.

Shoulders, M. D., & Raines, R. T. (2009). Collagen structure and stability. *Annual Review of Biochemistry, 78,* 929–958.

Statista. (2022). *Revenue of the global cosmetics market 2013–2026.* Statista Research Department.

Tan, C. C., Karim, A. A., & Latif, A. (2013). Extraction and characterization of pepsin-solubilized collagen from the body wall crown-of-thorns starfish (Acanthaster planci). *International Food Research Journal, 20,* 3013–3020.

Tang, P., Song, P., Peng, Z., Zhang, B., Gui, X., Wang, Y., Liao, X., Chen, Z., Zhang, Z., Fan, Y., Li, Z., Cen, Y., & Zhou, C. (2021). Chondrocyte-laden GelMA hydrogel combined with 3D printed PLA scaffolds for auricle regeneration. *Materials Science and Engineering C, 130,* 112423.

The Lancet Infectious Diseases. (2022). Time for Africa to future-proof, starting with COVID-19. *The Lancet Infectious Disease, 22*(2), 151. ISSN: 1473-3099.

Thomson, H. E. C., & Castilho, J. A. (2010). *Process of producing tomato paste*. United Stae Patent US20100104728A1.

Tyavambiza, C., Dube, P., Goboza, M., Meyer, S., Madiehe, M. A., & Meyer, M. (2021). Wound healing activities and potential of selected African medicinal plants and their synthesized biogenic nanoparticles. *Plants, 10*(12), 2635.

Uetani, K., & Yano, H. (2011). Nanofibrillation of wood pulp using a high-speed blender. *Biomacromolecules, 12*(2), 348–353.

Unilever. (2021). *2021 annual report and financial statement*. Unilever Nigeria PLC.

Walker, D. R., Hulgan, S. A. H., Peterson, C. M., Li, I., Gonzalez, K. J., & Hartgerink, J. D. (2021). Predicting the stability of homotrimeric and heterotrimeric collagen helices. *Nature Chemistry, 13*, 260–269.

Wang, S., Wang, F., Zhao, X., Yang, F., Xu, Y., Yan, F., Xia, D., & Liu, Y. (2022). The effect of near-infrared light-assisted photothermal therapy combined with polymer materials on promoting bone regeneration: A systematic review. *Materials and Design, 217*, 110621.

WTO (World Trade Organisation). (2020). Communication on trade in plastics, sustainability and development by the United Nations Conference on Trade and Development (UNCTAD). Committee on Trade and Environment. JOB/TE/63: 1–11.

Xu, S., Sang, L., Zhang, Y., Wang, X., & Li, X. (2013). Biological evaluation of human hair keratin scaffolds for skin wound repair and regeneration. *Materials Science and Engineering C, 33*(2), 648–655.

Yang, L., Guo, J., Zhang, S., & Gong, Y. (2017). Preparation and characterization of novel super-artificial hair fiber based on biomass materials. *International Journal of Biological Macromolecules, 99*, 166–172.

Yavuz, B., Chambre, L., Harrington, K., Kluge, J., Valenti, L., & Kaplan, D. L. (2020). Silk fibroin microneedle patches for the sustained release of levonorgestrel. *ACS Applied Biomaterials, 3*(8), 5375–5382.

Zoungranan, Y., Lynda, E., Dobi-Brice, K. K., Tchirioua, E., Bakary, C., & Yannick, D. D. (2020). Influence of natural factors on the biodegradation of simple and composite bioplastics based on casava starch and corn starch. *Journal of Environmental Chemical Engineering, 8*(5), 104396.

Chapter 8
Plastics and Polymer Manufacturing and Processing in Africa Today

Abstract The polymer industry belongs to the era of the modern industry far from the man or cattle-powered industry of previous era. Manufacturing in Sub-Saharan Africa is said to be lower than in all the other parts of the world. Much of the African output comes from the agricultural and extractive sectors. Plastic consumption in the region has reportedly risen by 150% in the past 6 years. Many of the companies involved in the manufacturing of plastics in Africa mainly carry out the conversion and molding of plastic resin into finished products and parts. While seven OPEC countries are located in the continent and around 8% of global petrochemicals production goes into plastics production, much of the resins used in plastic and polymer product manufacturing in Africa are imported. Rubber and cotton are the two main polymeric raw materials produced in the region. Others include paper, wood, leather, and polymers in food products, such that much of the polymer production within the content originates from agriculture and forestry.

Keywords Manufacturing · Plastic industry · Polymer processing · Injection molding · Polyurethane · Polymer technology

An Overview of Africa's Manufacturing Sector Today

Manufacturing in Sub-Saharan Africa is said to be lower than in all the other parts of the world. Much of the African output comes from the agricultural and extractive sectors (Austin et al., 2017). This has been attributed to various factors. One of which is the successive leadership's absence of capacity and/or will to develop infrastructure and sustain policies that promote manufacturing in the region.

Apart from the growth in manufacturing output that was experienced by some African countries between the period of 1924–1978 with most growth during this period experienced in South Africa, the manufacturing sector in most African countries hasn't seen much growth. During this period where manufacturing saw some growth in Africa some countries in Africa had up to 5% average yearly growth within a 10 year period. One perspective shared by Austin et al. (2017) is that the manufacturing industry of Africa should be characterized in its own unique way using better-suited criteria to measure progress and advancement beyond manufacturing output.

The plastic industry can be classified as a modern industry that makes use of mainly inanimate sources of energy. Prior to the modern industry, much of the industry was powered by humans and or draught animals. For example, an earlier chapter in the book discussed the use of porters to headload harvested resources such as rubber from the interiors to the coastal regions to be shipped out to Europe or the use of cattles by herdsmen and merchants to convey load across distances.

Countries like Kenya and Rwanda have focused more on their tertiary sector. Factors that make the tertiary sector more attractive to the government of such countries include, for example, Rwanda being a land-locked country that makes it more difficult to transport raw materials and finished goods across the borders. Focusing on the service sector with a focus on tourism generates income from international visitors who fly in and engage in tourism activities in the country.

Much of the synthetic plastics and polymer technologies have come from outside Africa. Although much of the industry in African countries has been based on the extraction and cultivation of the land some processing of polymer-based materials such as fiber spinning and weaving, leather processing, rubber tapping, gum extraction, silk extraction, and processing of polysaccharide materials such as starch are important industrial activities too.

Looking at the higher youth population compared to other more developed countries and the relatively less advanced technology. Rather than push for highly mechanized or automated development of the manufacturing sector in African countries, building the manual skilled labor force and human capacity might be a better alternative and more sustainable approach to growing the manufacturing sector in the region. For example, the Department of Trade, Industry and Competition of South Africa (DTIC SA) states that raw material price, less advanced manufacturing techniques, and research and development as some of the constraints to the South African plastics industry (DTIC SA, 2022).

Rather than aim for growing manufacturing output to meet the export demand and quality, an alternative approach is to grow sufficient manual skilled labor to achieve production output that meets the demand of the population. Excess can then be exported to other African countries and then the rest of the world. Applying this approach to the plastic industry would imply developing the mechanical and chemical plastics recycling as well as the biopolymers industry to meet the demands of the population. Hence, growing a plastics and polymer industry that builds on making optimal use of the already existing plastics and polymers accumulated in the environment. This offers the possibility of the African plastics and polymer industry emerging through a more sustainable path.

Plastic Manufacturers in Africa

African regions are now the fastest-growing market for plastics. These are goods made of plastics and machinery for processing plastics. The plastic use rate is reported to have risen by 150% within 6 years. Taking the example case of South Africa, 52% of the South African plastic industry, for example, comprises packaging plastics, while 13% comprises construction plastics and agriculture makes up 8% (DTIC SA, 2022). Of the packaging plastics, 29% are rigid and 23% are flexible plastic packaging. Other plastics are in the automotive and transport (6%), electric and electronics (6%), Furniture (2%), domestic and houseware (4%), mining and engineering 4%. Other plastic uses not within the aforementioned categories make up 5% of the South African plastics market. The plastic industry contributes 2.1% to the GDP and 21.8% to the GDP of the manufacturing sector (DTIC SA, 2020). The DTIC SA reports 1800 plastic companies as of 2022 and the plastic sector in South Africa employs around 60,000 workers.

An online search of some plastic companies within Africa suggests that many of the companies involved in the production of plastic goods make use of imported plastic resins. Much of the machines used are also imported. These include injection molding machines, extruders, blow molding machines, and thermoforming machines among others. Table 8.1 lists some plastic companies located in Africa. Where the company does not state the process used in the production of the products, knowledge of plastics and polymer processing techniques is used to determine the type of process and types of equipment used in its production.

The information for Vita foam, Nigeria was based on a visit to the company in 2018, which included a tour of the polyurethane foam production facilities. The company specializes in the production of polyurethane matrasses of various types. They also produce other polyurethane products such as furniture and footwear. The foam manufacturing takes place in the factories within Nigeria while the raw materials used in the production of polyurethane are imported. The company actively seeks to find alternative raw materials, which can be produced within the country. The company was established in 1962 and was listed on the Nigerian Stock Exchange in 1978. It supplies its products across Nigeria and to other countries in the West African sub-region.

Plastic and Polymer Regulatory Bodies, Associations, and Events within Africa

Across Africa, there are various plastic associations for manufacturers, consumers, and processors. There are also the government and non-government regulatory and standards bodies and agencies. Many countries across Africa have a Ministry dedicated to the environment such as the Ministry of Environment and Forestry. At the global level, African countries are also part of the UN Environment Assembly. Kenya

Table 8.1 List of some plastic manufacturing companies operating within the African continent

Plastic/polymer company	Country of operation	Product manufactured/supplied	Year established in Africa
Papilon Plastics	Nigeria	Planters/flower pots, furniture, baby chair potty, baby bath sets, mugs, cups, mop buckets	1987
Qualiplast Ltd	Ghanna	Buckets, basins, crates, baskets, jars, jugs, bottles, jerry cans, utensils, cutlery, food containers, waste bin, furniture…	1973
Malplast	Kenya	Blow molded PET and HDPE bottles	2004
Vita foam	Nigeria	Polyurethane faoms	1962
Kgalagadi Plastic Industry Ltd	Botswana	Flexible packaging of LDPE, HDPE, BOPP/PET laminates, and barrier films	1982
Soge plast	Angola	Household appliances, kitchen appliances, furniture	2001
Vector Engineering	South Africa	Prototype/custom injection molding and mold design for precision engineering parts	2015
Noel and Marquet (NMC)	South Africa	Polystyrene and polyurethane skirtings, cornicles, and wall panels (Imported from Belgium)	1950 (Belgium) 1999 (South Africa)

is emerging as a leader in global action to tackle the plastic waste crisis (UNEP, 2021). Kenya currently hosts the UN Environment program at its center in Nairobi.

Here, we look at some of them in different countries across Africa. In Kenya, there is the Kenya Association of Manufacturers established in 1959. The association recently launched the Kenya plastic action plan. Kenya also has the Green University Initiatives where member universities commit to greening their university campus and offering environmental courses in the sciences, management, and social sciences.

Countries in Africa such as Nigeria, Angola, Algeria, Benin, Botswana, Burkina Faso, Cabo Verde, Cameroon, Central African Republic, Chad, and other African countries are included in the 189 participants of the Basel Convention, Nigeria is one of the 53 signatories to the convention (United Nations Treaty Collection, 1989).

In Nigeria NESRA, the National Environmental Standards and Regulations Enforcement Agency (NESRA) exists for the purpose of developing specialized regulations regarding plastic waste at the national level. Another regulatory body within the country, the Standards Organization of Nigeria SON (Emodi et al., 2014) within its activities include setting the limits on plastic waste contamination. The organization also sets the standards for plastic product manufacturing in the country. Plastic-focused events are also held in Nigeria. An example is Plastprintpack, an annual international trade show that comprises exhibitions from various plastic manufacturers and suppliers as well as conferences (PlastPrintPak Nigeria, 2021). The trade show has also been held in other countries across Africa like Ethiopia, Ghana, and Kenya.

PlasticsSA in South Africa seeks to promote the growth of the South African plastics industry through offering industry training programs, introducing initiatives to address plastic pollution in the environment, advocating for sustainable practices in the plastic industry, and extending producer responsibility (PlasticSA, 2021).

Uganda's Coalition for sustainable development is dedicated to promoting actions to prevent climate change and meet sustainable development goals. Uganda is also a part of the clean sea campaign launched by the united nations environmental program (UNEP) in 2017. It is among the 63 partner countries in the program. 11 other African countries in the clean sea campaign are Madagascar, South Africa, Seychelles, Kenya, Uganda, Sudan, Nigeria, Benin, Ghana, Cote d'Ivoire, and Sierra Leone. The clean sea campaign also includes individuals, businesses, sports stars, and other groups. For example, more than 63 hotels are also members of the UNEP clean sea campaign (UNEP, 2017). These are largely waterside hotels that commit to terms such as avoiding the use of plastics in the hotels.

At the sub-regional level, the East African Community Legislative Assembly has various committees dedicated to meeting and addressing various issues within the member countries. One such committee is the Committee on agriculture, tourism, and natural resources. The management of production and use of plastic is among the areas of interest of the committee (EAC, 2017). An example of the plastic-related issue that was addressed by the committee was the bill on polyethylene bag production and use within the member countries, the EAC Polyethylene Materials Control Bill 2016. Tanzania, Rwanda, Kenya, Uganda, and Burundi are the member countries of the EAC.

In the textiles industries where polymeric materials such as cotton, silk, and polyester are widely used, various organizations have been established to facilitate, regulate, or promote the trade of textile and garments at national, regional, or international levels. Examples of such are the African growth and opportunities act. Established in 2000, the act allows selected African countries to enjoy duty-free imports of selected goods into the United States. 34 countries in Africa have been declared eligible for AGOA (TPCC, 2016). Members of AGOA who have taken advantage of the act include Kenya, Lesotho, Namibia, Swaziland, Madagascar, Mauritius, Tanzania, Nigeria, and South Africa (World Bank, 2018). There was also the multi-fiber agreement (MFA) by the world trade organization, which has been phased out in 2005 (Wijayasiri & Dissanayake, 2009). The arrangement placed a quota on trade between developed and developing countries.

Rubber and Rubber Products' Production in Africa

Liberia, Nigeria, Zaire, Ivory Coast, Cameroon, Central African Empire, Ghana, Mali, and Congo, are the top producers of rubber in the African region. In 2018, Africa collectively produced close to 817 thousand tonnes of natural rubber according to FAOSTAT. Asia, the largest producer of natural rubber produced 12.8 million tonnes in the same year. Asia has always produced more rubber than Africa. In 1961, the figures reported for Africa and Asia by FAOSTAT were 147,743 tonnes and 1,936,370 tonnes, respectively. No data were reported for the production of synthetic rubber in Africa. However, Zimbabwe reported 4520 metric tonnes in 2009 dropping to 70 tonnes by 2013. Between 2009 and 2013, an average of 1,510 tonnes of synthetic rubber production was reported for Zimbabwe. An average of 66,410 tonnes of synthetic rubber production was reported for South Africa between 1995 and 2008, showing an overall increase from around 56 thousand tonnes in 1995 to just over 75 thousand tonnes in 2018.

One of the multinational companies that produce rubber and rubber products in Africa is OLAM. It is a global food and agriculture business with locations across 15 countries in Africa and also others located in countries in the Asia Pacific, Europe, Latin America, and North America. Its polymer and polymer-derived products include rubber, cotton, wood, proteins such as pulses, and starch products such as grains and rice (OLAM, 2020). It has large estates for the cultivation of rubber across Africa.

Some examples of rubber product manufacturers in Africa include Allan Maskew, a subsidiary of the Argent group (Argent, 2021), and ERM rubber moldings. According to the UN data in 2007, South Africa had 500 rubber establishments, and Tunisia had 1009. In 2011, 455 South African establishments were reported in the database; however, no data were given on Tunisia. In 2007, Mauritius reported 83 rubber establishments while Morocco reported 26 rubber establishments. Senegal, Eritrea, Ethiopia, Malawi, and the United Republic of Tanzania reported between 1 and 4 rubber establishments in 2007. These figures are only based on data that were available to the UN at the time of the study. Production records show that despite the low or unreported number of establishments in this database rubber production is active in several countries in Africa.

According to the UN data in 2018, the top natural rubber-producing countries in Africa are Cote d'Ivoire, Nigeria, Liberia, Cameroon, Gabon, Ghana, Guinea, Democratic Republic of Congo, Congo, and the Central African Republic. Table 8.2 lists the production quantities for each of these countries for 2018 (UN data, 2018). The top two rubber producers Cote d'Ivoire and Nigeria produce more natural rubber than the countries in Middle Africa (99,000 tonnes). The figures provided are given to the nearest thousand tonnes.

Table 8.2 Rubber production in some African countries

Country	Production thousand tonnes
Cote d'Ivoire	461
Nigeria	145
Liberia	71
Cameroon	55
Gabon	25
Ghana	24
Guinea	17
Democratic Republic of Congo	15
Congo	3
Central African Republic	1

Textile Manufacturers in Africa

Kenya and Lesotho are the top African exporters of textiles to the United States today. South Africa, MauritDemocratic Republic of Congoius, Zambia, and Madagascar are also other textile manufacturing countries across the Congocontinent. The textile industry makes use of natural and synthetic polymers such as cotton and polyester and silk. In Lesotho, for example, textiles and apparel form a large part of the manufacturing activity in the country. 57% of its export to its neighboring country South Africa comprises textiles and apparel (World Bank, 2018). Textiles and apparel and wholesale and retail industry in Lesotho provide 84% of employment in the country. Although it only makes about 6% of the revenue generation due to the low-value addition. Lesotho's modern textiles and apparel industry began its growth in the late 1980s with the arrival of foreign investors from South Africa and Taiwan. Many of these investors left South Africa due to the unfavorable conditions for manufacturing during the apartheid era. The growth of the textiles industry was also influenced by the African growth and opportunities act AGOA to which it became a beneficiary in 2000. Through AGOA, it enjoyed duty-free export of selected goods to the United States. Although there was a reduction from 94 to 61% in export to the United States between 2005 and 2016 (World Bank, 2018) partly owing to the anticipated expiration of AGOA, the act has since been extended to 2025.

Countries that benefit from AGOA export around 83% of the cotton produced, mainly to the United States (World Bank, 2018). With the exception of Nigeria where 80% of the cotton production is consumed within the country (Owen et al., 2016), other large cotton producers; Cameroon, Mali, and Cote d'Ivore all export most of the cotton produced to other countries. The yarns used in textile and apparel in the AGOA countries are mostly imported from the EU, US, and China. Regional trading of clothing and textiles between African countries also occurs. For example, South Africa, Namibia, Mozambique, and Zambia export clothing to other countries in the African region. The textile's market involvement also varies at different stages.

Within Africa, spinning is likely done in Lesotho, Zambia. Mozambique, Botswana. Mauritius and South Africa are potential locations for product design and marketing according to the World Bank Report (World Bank, 2007).

Textile factories based on Lesotho produce for brands that include Wrangler Corporation, Mr. Price, Levi Strauss, Woolworths, Walmart, Costco, Foschini Group, Pick n Pay, JC Penny, Truworths, and Pep Group (World Bank, 2018). Most of the firms concentrate on textiles and apparel for the fast fashion industry. No known manufacturers produce luxury brands or high-value apparel such as coats, underwear, or formal suits.

Around 98% of the cotton produced in Africa is exported as raw cotton. Meanwhile, Africa imports a large percentage of its textiles. The African Fabric Holdings, a subsidiary of four brands namely; Vlisco, Uniwax, GTP, and Woodin 2018 reported a net turnover of 231.9 million Euros. This was a rise from 228.3 million Euros in the previous year. The main target market for the company is Sub-Saharan Africa (Africa Fabrics Holdings B.V., 2018). Its activities in the textiles industry are in designing, manufacturing, and selling dyed and printed fabrics. These prints have become very common in West Africa and central Africa. The company has production factories in Netherland, Ghana, and the Ivory Coast.

Other Polymers

In Ethiopia, the number of firms in the rubber and plastics industry rose from 15 to 186 between 1996 and 2016. Other polymer-related industries also saw growth within the period stated. The paper industry had 24 firms in 2016, a rise from 5 firms in 1996. Wood rose from 25 in 1996 to 64 in 2016. The textile industry in Ethiopia saw a jump from 32 textiles firms in 1996 to 258 by 2016. Leather only added three more firms between 1996 and 2016 going from 63 to 66 firms within the stated period. Food and beverages that are assumed to include polymers such as polysaccharides and proteins rose significantly from 160 in 1996 to 668 firms by 2016 (Jones et al., 2019). The data on the number of firms in polymer-related industries in Ethiopia, Tanzania, and Cote d'Ivoire are summarized in Table 8.3. Proteins used for food, cosmetics, pharmaceutical, and other applications are also relevant polymers to consider. Lesotho, for example, exports trout to Japan and South Africa in amounts above that of other suppliers (World Bank, 2018).

Challenges and Future Prospects of Plastics and Polymer Industry in Africa

Some of the challenges to the plastics and polymer industry in Africa from information gathered across different sources as discussed within the text include lack

Table 8.3 Number of firms in polymer-related industries In Ethiopia, Tanzania, and Cote d'Ivoire

Industry	Ethiopia		Tanzania		Cote d'Ivore	
	1996	2016	2008	2013	2003	2014
Textiles	32	256	40	45	9	12
Rubber and plastics	15	186	15	43	41	55
Leather	63	66	3	2	12	35
Wood	26	64	17	68	54	69
Paper	5	24	7	14	5	4
Food and beverages	160	668	236	430	63	168

of specialized skills, weak backward linkage, lack of water infrastructures, lack of secure industrial housing infrastructure, expensive and ineffective internet service, access to finance and issues with internationally recognized standards certification.

One of the challenges to the plastics and polymer industry is the competition African manufacturers face with cheap imports. China is Africa's current top trade partner and imports goods worth billions of dollars to different countries in Africa every year. Analysis in recent studies shows that the import of textiles from China into Africa is having a negative impact on the textiles industry in the African countries to which these textiles are imported (He, 2020).

With the increasing need to manage the plastic pollution crisis, companies can embrace the idea of using locally available recycled plastic waste as their main feedstock. This will reduce the dependency on imported virgin resins for plastic product manufacturing. The limitation of this is the challenge of using recycled plastic in products such as food packaging, pharmaceutical, and personal care products. Advanced recycling technologies are required to recycle plastics to achieve the level of purity required for such applications. Nonetheless, other products can be produced with recycled plastics using standard techniques to boost the output of the plastic industry and also utilize the existing plastic waste resource.

Conclusion

Much of the production output from the continent is based on the primary sector. Cotton, rubber, wood, leather, paper, and food are some of the polymer-based product output from the African region. Cote d'Ivoire and Nigeria lead in rubber production while Ethiopia and Tanzania lead in textiles. Seven African countries are among the 13 member countries of OPEC. Petroleum still serves as the main source of synthetic plastics and polymers today. However, products manufactured from modern synthetic plastics require imported resins as feedstock. Replacing the products made from these synthetic plastics with those readily available feedstock like cotton can reduce dependency on imported synthetic polymer resin. Existing waste plastics can also be mined from the environment to serve as feedstock for new plastic products while simultaneously addressing the issue of pollution.

References

African Fabrics Holdings B.V. (2018). *Annual report for the year ended 31 December 2018.*

Argent Industrial Limited. (2021). *2021 annual report.*

Austin, G., Frankema, E., & Jerven, M. (2017). Patterns of manufacturing growth n sub-Saharan Africa: From colonization to the present. In K. Hjortshoj, H. O'Rourke & J. G. Williamson (Eds.), *The spread of modern industry to the periphery since 1871.* Oxford University Press.

DTIC SA Department of Trade, Industry and Competition South Africa. (2022). *South African Plastics Industry.* Retrieved March 06, 2022, from http://www.thedtic.gov.za/sectors-and-services-2/industrial-development/plastics/

East African Community East African Legislative Assembly. (2017). *Report of the Committee on Agriculture, tourism and natural resources on the oversight activity on "waste management in East African community partner status" 10th–14th April 2017.*

Emodi, V. N., Yusuf, S. D., & Boo, K. J. (2014). The necessity of the development of standards for renewable energy technologies in Nigeria. *Smart Grid and Renewable Energy, 5*(11), 259–274.

He, Y. (2020). The Impact of imports from China on African textile exports. *Journal of African Trade, 7*(1–2), 60–68.

Jones, P., Lartey, E. K. K., Mengistae, T., & Zeufack, A. (2019). Sources of manufacturing productivity growth in Africa. In *World Bank group: Africa region office of the chief economist.* Policy Research Working Paper 8980.

Olam International Limited. (2020). *Anual report: Unlocking long-term value and driving sustainable growth.*

Owen, M. M., Ogunleye, C. O., & Orekoya, E. O. (2016). The Nigerian textile industry: An overview. *Nigeran Journal of Polymer Science and Technology, 11,* 99–108.

PlasticSA. (2021). *Plastics, people and planet: Anual review 2020/2021.*

PlastprintPak Nigeria. (2021). *Post-show report.* 26–28 October 2021 Landmark Centre—Lagos. Retrieved February 20, 2022, from www.ppp-nigeria.com

TPCC Trade Promotion Coordinating Committee of the United States. (2016). *National export strategy 2016.* International Trade Administration.

UNEP. (2017). *Over 65 Phuket hotels join UN Environment #CleanSeas campaign.* Press Release 14 December 2017.

UNEP. (2021, February 18). Kenya emerges as a leader in the fight against plastic pollution. *Nature Action.* Retrieved February 02, 2022, from https://www.unep.org/news-and-stories/story/kenya-emerges-leader-fight-against-plastic-pollution

United Nations Treaty Collection. (1989). Basel Convention on the control of transboundary movement of hazardous wastes and their disposal. Basel. 25(1) Vol. 1673 p. 57. Status as of 20 February 2022

Wijayasiri, J., & Dissanayake, J. (2009). The ending of the multi-fiber agreement and innovation in the Sri Lanka textile and clothing industry. *OECD Journal, 4,* 157–188.

World Bank. (2007). *Vertical and regional integration to promote African textiles and clothing exports—A close-knit family?* (Report No. 39994-AFR).

World Bank. (2018). *Unlocking the potential of Lesotho's private sector: A focus on apparel, horticulture, and ICT* (pp. 21–36).

Chapter 9
Chemical Recycling and Energy Recovery from Plastics and Other Polymers in Africa

Abstract Previous chapters of this book have discussed the origins of plastic in Africa, plastic production in Africa, and the mechanical recycling of plastics within the continent. From the discussion so far, we see that, despite some countries implementing bans and other measures to reduce plastic waste accumulation, there is still an increasing trend in plastic production in Africa and across the world. This chapter reviews chemical plastic recycling in Africa. The first few sections discuss chemical recycling and energy recovery technologies. The later part of the chapter discusses some of the chemical recycling and energy recovery projects in some countries across Africa. In doing so, it gives the reader an idea of the current status and future potential of chemical recycling and energy recovery in terms of the available resource and potential output and impact within the African continent. The polymer wastes considered here are largely plastic from municipal solid waste. That is waste generated from the daily activities of homes, businesses, and other commercial activities. It excludes the plastics scraps from the manufacturing of plastics, and the conversion or processing of plastics during the manufacturing of plastic products. A section considered the anaerobic digestion of waste from livestock processing which comprises mainly proteins and lipids.

Keywords Chemical recycling · Pyrolysis · Gasification · Energy recovery · Polycondensation · Depolymerization

Plastics Recycling Processes

Municipal solid waste in general comprises compostable, combustible, and dormant parts (Simegn et al., 2021). The compostable parts are mainly food and other organic wastes that are broken down back microbes in a compost pile. The combustible parts are the parts that have relatively higher calorific values and can be burnt to generate heat. These include plastics, paper, and other materials, while the dormant part of the municipal solid waste is those parts that are not broken down by burning or microbes. Examples are stones ceramic and some metals. This chapter focuses mainly on the plastic component of municipal solid waste. The chemical recycling and energy recovery process discussed herein also applies to other polymeric materials like paper, polymers in food, cellulosic fibers, collagen, and other polymers.

To begin with, this section reviews the different plastic recycling processes. These are generally divided into four categories; Primary, secondary, tertiary, and quaternary plastic recycling. Primary recycling is also referred to as in-plant recycling. During the processing of plastic reins/pellets into other products using processes such as injection molding, extrusion, and blow molding, some parts may be rejected due to process inefficiencies or errors. Scraps are also generated in, for example, the runners and sprues in injection molding or the trimmings from blow molding. These can be reprocessed and put back into the system to reform new products. These are the primary or in-plant recycling.

Secondary recycling refers to the mechanical recycling of typical thermoplastics by physical size reduction followed by thermal deformation and reforming into other products. The plastics retain the chemical structure throughout. This is similar to primary recycling except it refers to post-consumer plastics. These are plastics that have already gone through at least one usage life span.

Tertiary recycling refers to chemical recycling or feedstock recycling. As these names imply, in this type of recycling, the plastics are chemically broken down into their monomers or other smaller molecules such as carbon dioxide, hydrogen, methane, and water. These compounds can then be used as raw materials for the production of either new virgin plastics or other chemical products.

Quarternary recycling refers to recycling where the plastic is converted into energy. This is mainly by incineration, an exothermic process that releases the stored chemical energy in the polymers. The energy is usually obtained in the form of direct heating or used to power a turbine to generate energy. Figure 9.1 summarizes the different types of recycling.

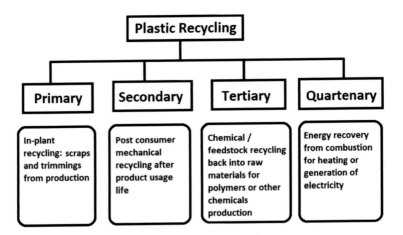

Fig. 9.1 Different types of recycling summarized

Chemical Plastics Recycling Technology

These two processes are looked separately here because so far primary and secondary recycling have been more widely utilized and fully commercialized despite the relatively lower global recycling rate (~9%) compared to landfilling rate (~79%) (Greyer et al., 2017).

Although the plastics with resin numbers 1–6 are generally regarded as recyclable. Mechanical plastics recycling comes with the limitation of the progressive deterioration of the plastic quality with each recycling. This means that in reality, most plastics can only be effectively recycled once. Subsequent recycling will result in the loss of the properties of the plastics. This is a result of the weakening of the bonds between the chains and molecules that make up the polymer. This weakening of the bonds results in alteration of the primary and secondary structure of the polymer to which its properties are attributed.

Thermoset plastics have the advantage of fast low-temperature processing. Thermoset polymers have generally been regarded as nonrecyclable. This is because they do not have a melting point, rather they degrade beyond their tolerable temperature. Therefore, thermosets are not included in the resin identification numbers 1–7. Although the recycling of certain plastics has not been commercialized, chemical recycling not only offers the possibility of endless recycling of plastics but also of a wider range of plastics which could include thermosets. Thermosets such as epoxies, polyurethanes, and phenol-formaldehyde, thermoplastics like polystyrene, polycarbonates, and polypropylene as well as nonpolymeric organic wastes such as carbon fiber and cellulosic materials, can all be chemically recycled. The chemical recycling of thermoset plastics is much less explored compared to that of thermoplastics.

Polyethylene is the most consumed plastic globally. An estimated 49 million tonnes of polyethylene was consumed in the world in 2021. This is projected to rise to 50 million tonnes by 2022 (Statista, 2021a, b). The second highest plastic consumed globally is polypropylene and in 2017 172 million tonnes of polyethylene and polypropylene were produced globally (Statista, 2021b). Therefore, there have been a lot of studies into the recycling of these two commodity plastics.

While mechanical recycling does not result in alteration of the chemical structure of the polymer, chemical recycling on the other hand is targeted at breaking down the polymer structure into smaller molecules. Depending on the type of chemical recycling, this can be into monomers or smaller molecules like carbon dioxide, hydrogen, and water.

The current challenge with chemical plastics recycling lies in safety, energy efficiency, and large-scale production. Like mechanical recycling, it is also affected by the challenge of collection of post-consumer plastics, especially those that have ended up in hard-to-reach places like the sea, drainage systems, and bushes. The collection and pretreatment of these plastic wastes add to the time and cost of recycling. The different types of chemical recycling are summarized in Fig. 9.2.

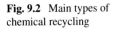

Fig. 9.2 Main types of
chemical recycling

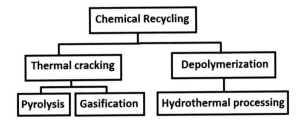

Hydrothermal Processing

While hydrothermal processing is a technique that has been in use since the 1990s.
The term was introduced in the late 1800s by the geologist Sir Roderick Murchison.
In nature, it refers to the different formations of rocks and minerals which occur as
a result of the action of water at elevated temperatures and water within the earth's
crust.

The temperatures for hydrothermal processing are within hundreds of degrees
Celcius and pressure well above atmospheric temperatures. For example, in one
of the early applications of the processes for the formation of barium carbonate and
strontium carbonate crystals in 1839, Robert Bunsen applied a temperature of 200 °C
and a pressure of 100 bar (Yoshimura & Byrappa, 2008).

Although hydrothermal processing primarily makes use of water, variations of the
process make use of other non-aqueous solvents like alcohols and ethylene glycol.
The addition of catalysts and alkali are also included to enhance the process and/or to
direct the reaction towards the production of specific desired products. For example,
hydrothermal leaching of bauxite in the production of aluminum hydroxide makes
use of sodium hydroxide (Zong et al., 2019). PET has also been reformed into
hydrogen and other fuels using catalytic reforming in phenol as solvent using Ni–
Pf/Al catalysts on titanium nanoparticles (Nabgan et al., 2020). Although steam was
still used in the process.

Because it makes use of main water, hydrothermal processing is often preferred
as a more environmentally benign alternative to the processes requiring more toxic
solvents. Water is abundant, it makes up 73% of the earth. It is non-hazardous and
non-toxic and can exist in a liquid, solid, or gaseous state. The physical state of water
can be altered by manipulating the temperature and pressure. At a temperature of
25 °C and a pressure of 0.1 MPa, water exists as a liquid on earth. A significant
phenomenon of water for hydrothermal application is the changes in its solvent
properties at water's supercritical condition. At the supercritical state, water can
simultaneously exist as liquid, solid, and gas. Under this condition, water behaves
like an organic solvent like methylene chloride and becomes soluble in organic
substances that under standard conditions are not soluble in water. This phenomenon
is attributed to the reduction in the dipole moment of water under the supercritical
condition. It also releases more hydrogen and hydroxyl ions and becomes more
reactive at this supercritical condition thus boosting the ability of water to serve as a

catalyst for hydrolysis processes. Therefore, in the hydrothermal treatment process, water simultaneously acts as a reactant as well as a catalyst and a solvent.

The low viscosity of water allows for a fast diffusion rate and, under such high pressure, the collision rate is faster. This removes limitations to mass transfer thus allowing for faster and more probable reactions. Variations of hydrothermal processing include hydrothermal extraction, hydrothermal decomposition, and hydrothermal sintering among others.

Pyrolysis

This process is a chemical recycling process used to obtain solid residue in the form of char, liquid oil, and gases from the degradation of polymeric materials. The volatile fraction is made up of condensable and noncondensable gases with high calorific values. It occurs between 350 and 900 °C under anaerobic conditions. On average, 1 kg of a mixture of plastic waste will yield 0.8 kg of pyrolytic oil. This has an energy potential of 39.6 MJ per kg on average (Rehan et al., 2017). These values vary for different processes and also depend on the efficiency of the operation.

The products of pyrolysis are affected by the temperature, feed composition, and process time. Therefore, these parameters can be varied to control the product of the pyrolysis process. Generally, higher temperature favors the production of a more volatile fraction which is more desirable for use as fuel.

Pyrolysis of thermoplastics and thermosets has been explored. In one study, the pyrolysis of ortho-phthalic polyester reinforced with sheet molding compound of fiber glass carried out under laboratory conditions at temperatures varying between 300 and 700 °C over a duration of 30 min yielded 72–82 wt% of solid residues, 9–13 wt% liquid residues, and 6–12 wt% gas residues (Torres et al., 2000). It was observed that the optimal temperature was 400 °C and the liquid comprises a complex mixture of non-toxic C_5–C_{20} hydrocarbons which mainly had aromatic structures with gross calorific values between 34 and 37 MJ/kg. The gas fraction is composed mainly of carbon monoxide and carbon dioxide with low calorific values between 13.9 and 16.4 MJ/m^3 N.

In another study on the pyrolysis of a thermoplastic, polyethylene, the pyrolysis was carried out at a temperature ranging between 500 and 600 °C over a duration of 12.4–20.4 s in a fluidized bed reactor (Zhao et al., 2020). The process yielded 81.2–28.5 wt% pyrolysis liquid composed mostly of mono-olefins, cyclic alkanes, and dienes. with a reduction in yield when the temperature was increased from 500 to 600 °C. The gas phase was between 8.2 and 56.8 wt% for the temperature range tested, respectively, consisting of paraffins, olefins, hydrogen, and 1,3-butadiene.

Pyrolysis can be carried out on a mixture of plastics of different types as well as mixtures of plastics with other materials. Thermosets can also be chemically recycled using pyrolysis. This includes thermosets with reinforcements such as glass fiber and carbon fiber (Ginder & Ozcan, 2019). This technology is established to the point of commercialization at a small scale (Oliveux et al., 2015) in advanced economies.

Pyrolysis and other chemical recycling technologies are less advanced in African countries. This can be set as an advantage by deriving lessons from tried and tested methods from more advanced technologies and then developing the chemical recycling industry of African countries based on the most well-established technologies.

Gasification

This is also a chemical recycling process, however, it occurs at a higher temperature than pyrolysis and it occurs under aerobic conditions. Process temperature is usually around 800 °C but can vary between 500 and 3000 °C (Seabea et al., 2019) and the main product gas is synthesis gas (syngas) and hydrogen (Lopez et al., 2018). It is a thermal cracking process that breaks down the polymers into smaller molecules. A gasifying agent is required and this can be air, oxygen, steam, or other. A review of literature on gasification suggests that air is more commonly used (Lopez et al., 2018). The type of gasifying agent used affects the product of the process. For example, the use of steam as a gasifying agent can result in a higher heating value and H_2 content of the product. This also results in higher tar content in the product which is undesirable.

Unlike pyrolysis, gasification occurs in the presence of air, and its main products are in the gaseous state with tar produced as a by-product. The gasification process has been in existence for decades and has been used for biomass and coal since the early nineteenth century (Pereira & Martins, 2017). Recent advancements in the technology include combined gasification of plastics alongside biomass and coal and optimization of process conditions to eliminate the production of tar. Different combination of plastics in gasification has also been studied to determine the best mixture of plastic waste in the gasifier to obtain the most desired output.

In an example of steam gasification of polypropylene, 1 g of polypropylene was fed into the gasifier with steam flowing in at 4.7 ml per hour (Wu & Williams, 2009). The gasifier was operated at 800 °C and a pressure of 1 atmosphere. The product gas contained 64% hydrogen, 25.7% carbon monoxide, 6.4% carbon dioxide, 3.3% methane, and 0.6% compounds with two carbons or more. In a simulated study of the gasification of polyethylene and polypropylene, it was observed that gasification of pure polyethylene and steam to feed mass ratio of 1.5 at a temperature of 900 °C achieves the highest synthesis gas production rate compared to either pure polypropylene or mixtures of polypropylene and polyethylene (Seabea et al., 2019). The study showed that the products of gasification are affected by operating parameters like temperature, feed to gasifying gas ratio, and the type of gasifying agent used. This is based on the understanding of the possible reactions taking place during gasification and that certain conditions support certain reactions over others. For example, a higher temperature favors endothermic reactions leading to increased hydrogen and carbon monoxide production. However, increased temperature results in reduced production of methane, carbon dioxide, and water since these reactions are not favored at higher temperatures.

The production of hydrogen gas from plastic waste has generated much interest in recent years. Simulated gasification processes show that hydrogen-rich syn gas is obtainable from high-temperature gasification of medical plastic waste using steel converter ash as a catalyst. (Qin et al., 2022). The use of gasification for medical plastic waste is particularly important. The high temperature is important in killing any potential pathogens, especially thermophiles which may survive conventional mechanical recycling and occurs at much lower temperatures. The gasification temperature for this process is between 900 and 1100 °C over Fe_2O_3 and CaO catalysts in the converter ash.

Energy Recovery Technology

Synthetic plastics are primarily derived from crude oil. Although today synthetic plastics are now being produced from more sustainable feedstocks such as used cooking oil (Moretti et al., 2020). Much of the commercial plastic production is based on crude oil. These plastics have significantly high amounts of chemical energy stored within them which can be recovered from their combustion. The energy is obtained in the form of thermal energy which can be used for direct heating or used to power steam turbines to generate electricity.

Other energy recovery technologies used to obtain fuel or energy from waste include anaerobic digestion of food waste and transesterification of waste oils. Briquettes for cooking and heating are produced from a mixture of components like cow dung, agricultural wastes, and cactus (Shuma & Madriya, 2019) Japan has implemented incineration into its municipal waste management system and has maintained a good reputation for its waste management effectiveness (UNEP, 2013).

Energy recovery from plastic and other polymeric waste involves the incineration of the waste within a controlled chamber. Well-controlled incineration ensures that the toxic products of the combustion of plastics are captured during the combustion and then properly disposed of. Research has been carried out to explore the recovery of metals from the incineration of waste and utilizing these metals (Bakalar et al., 2021).

A study on Medina city looked at the recovery of energy fuels from plastic waste and food waste through pyrolysis and anaerobic digestion, respectively. Food waste makes up an estimated 40% of municipal solid waste, while plastic waste makes up 20%. Based on the analysis carried out, an estimated 1409.63 TJ is recoverable from the food waste generated in the city and 5619.8 TJ of energy is recoverable from pyrolysis of all the plastic waste generated in the city of Medinah in Saudi Arabia. The plastic waste pyrolysis oil can potentially generate 58.81 MW of electricity for the city annually Rehan et al., 2017). Although there needs to be a careful assessment to consider, if incineration for energy recovery is a better option compared to other plastic waste management strategies like recycling (Ekwall et al., 2021). Such assessment should consider factors like the energy recoverable, the prospects of reusing, the energy input, and environmental impact over the life cycle of the products.

Note that although pyrolysis and gasification are classified under chemical recycling, the products from these chemical recycling processes such as methane and hydrogen can also be used as fuel for heating or the generation of electricity. The energy recovery process is placed in a different category here since it is used to directly generate energy. The plastics are directly used to generate heat through combustion. This heat can then be directly used to heat up processes or to power turbines to generate electricity.

With a population of well over a billion, generating around 125 million tonnes of waste per year (UNEP, 2018). The next sections take a look at specific case studies of some plastic and polymer recycling and energy recovery projects in some countries in the African continent.

Energy Recovery from Municipal Solid Waste in Ethiopia

So far, the waste-to-energy industry in Africa is still in its infancy. The first-ever waste-to-energy plant was built in Ethiopia in 2017. As of 2020, 0.7% of Ethiopia's electricity generation was from municipal solid waste (Desta et al., 2022). The rest of Ethiopia's electricity mix comes from hydropower (89.5%), wind power (7.6%), geothermal (0.2%), and diesel (92%). Ethiopia has the second-highest population in Africa, next to Nigeria.

The Reppie municipal waste to energy plant in Ethiopia Adis Ababa incinerates an estimated 1400 tonnes of feedstock daily. This amounts to about 100,000 tonnes of waste incinerated at the plant annually accounting for around half of the municipal solid waste generated in the city. Some of the challenges of waste incineration are the generation of fly ash from the process.

One of the by-products of chemical recycling through processes like pyrolysis (Pandey et al., 2020) and energy recovery (Simegn et al., 2021) processing of plastic waste is ash. Two types of ash are generated in the incineration of municipal solid waste; bottom ash and fly ash. The bottom ash is the portion of the ash that contains particles ranging between 10 and 15 mm particle size. Fly ash is the powdery portion of the ash that consists of finer particles. The latter is more easily dispersible in the air and can pose an environmental and health concern if not properly handled. Studies have been carried out to explore the utilization of fly ash from the Reppie incinerator in the production of cement (Simegn et al., 2021). Such use of the by-product from the incineration of municipal waste contributes towards addressing the negative aspect of waste incineration and reducing the amount of cement required in the production of concrete which comes with some adverse environmental impacts.

Energy Recovery from Municipal Solid Waste in Uganda

Although during the time of this publication, Uganda has no known public municipal solid waste to energy incinerator, there have been studies carried out to analyze the potential for such waste to energy facilities in the country. A study on municipal solid waste to energy was carried out in Kampala recently (Amulen et al., 2022). Of the 1.9 billion tonnes of municipal solid waste generated globally every year, the city of Kampala in Uganda generates around 350,000 tonnes. Simulated studies to analyze the amount of energy that could be generated from municipal solid waste in Kampala suggest that up to 774 kWh can be generated from municipal solid waste generated in Kampala, enough to meet the electricity needs of an estimated 1.62 middle-income earning homes in Uganda. Such a plant is estimated to incur a start-up capital of USD 157 million.

Recyclables which include plastics make up an estimated 15% of municipal solid waste found in landfills in Kampala, while 43% comprise organic waste and 42% comprise mixed fines (sometimes also classified as organic waste). The high composition of organic waste suggests that other waste energy recovery processes such as anaerobic digestion to obtain methane might be better than incineration (Amulen et al., 2022).

Blood, Rumen, and Bone Waste Anaerobic Digestion for Bioenergy Recovery

Meat is a major source of protein in many parts of the world. Meat is processed in abattoirs where livestock like cattle and goats are slaughtered and processed for sale as meat. In the slaughtering and processing of livestock, waste is generated. This includes blood, internal organs, hoofs, bone, and hair. These materials contain polymers in the form of collagen, keratin, and other proteins as well as lipids. All of these wastes are useful for different applications such as animal feed, fertilizer, and the production of glue. The anaerobic digestion of blood, rumen, and bone waste is included here because it involves the chemical breakdown of polymers within these materials into new compounds used as fuel and fertilizers.

To derive value from the waste requires the input of knowledge, skill, and additional resources. When the butchers and traders are not willing to dedicate the time and resources or lack the skill and knowledge to do so, the waste simply gets discarded. This is also down to the value of the consumed part of the meat compared to the value from the waste and the scale of operation.

In a study to explore the potential for energy generation from blood and rumen waste deposits in an abattoir in Shashemene city Ethiopia (Kefalew & Lami, 2021), it was observed that annually 283,605 kg of blood waste and 405,150 kg of rumen waste was generated in the abattoir. This was obtained from around 13,505 cattle slaughtered annually, estimating the average weight of the livestock slaughtered to be 250 kg. The blood and rumen are estimated to comprise 8.4% and 12% of the body weight, respectively. From this, it was possible to produce a total of 206,626.5 m^3 of biogas through anaerobic digestion. The study also estimated an annual waste of 688755 kg. This waste can potentially generate 371,927.7 kWh of electricity a year. The study also showed that around 43,184.9 kg dry weight of biofertilizer could be produced from the waste from the slaughterhouse. This can be obtained from the decomposition of the waste to produce plant nutrients for agriculture.

Other similar studies have been carried out in abattoirs in other cities in other African countries. A study of an abattoir in Hawassa city estimated an annual mass of waste generation of 885,881.6 kg (Sindibu et al., 2018). Another study in the abattoir located in Minna in Nigeria estimated 873,810 kg of waste produced annually from the slaughter waste (Ahaneku & Njemanze, 2015) with around 60 animals slaughtered daily (Audu et al., 2020). Other than the bioenergy and the biofertilizer derived from the blood and rumen waste in the abattoirs. The proper recycling and utilization of this waste reduce the greenhouse gas emission by 952.4 tonnes of CO_2 equivalent per year from the fossil fuels that would have been used instead (Kefalew & Lami, 2021).

To harness the potential energy and fertilizer from the meat slaughtered in abattoirs, the process of slaughtering livestock in the abattoirs should be better organized to ensure that the waste is effectively collected for processing into valuable products. Bioenergy and fertilizer producers should establish partnerships with abattoirs to collect and utilize the waste generated. Incentives should be created to encourage the collection and processing of slaughterhouse wastes.

With the world currently facing an increase in the price of fossil fuel and a scarcity of fertilizer, an alternative sustainable source of these valuable resources is important to energy supply and food security. For a country like Ethiopia with no known crude oil reserve, the ability to internally generate energy and fertilizer is of much importance to sustain its industry and population.

Feedstock and Energy Need in Africa

Many products that are essential for modern living today are derived from plastics and polymers. An estimated 8.3 billion tonnes of plastics have been produced since the introduction of plastics into commercial production around 1950 (Jambeck et al., 2018). One study estimates that 60–99 million metric tonnes of mismanaged plastic waste were produced in 2015 across the world (Lambreton & Andrady, 2019). At the current rate of increase in annual plastic waste generation, this could go up to 265 million tonnes annually by the year 2060.

It is estimated that if the current production rate persists around 20% of crude oil consumed in the world would be used for plastic production by 2050 (UNEP, 2018). The current estimate of global annual plastic production is 400 million tonnes. Of this, around 36% are single-use plastics used for packaging. Africa represents 1% of global production of single-use plastic production as of 2014. As of 2015, 47% of plastic waste generated was plastic packaging. While other plastics products such as textiles, materials used in building and construction, and others tend to have a longer usage lifespan, single-use plastic packaging has a very short useful life span and tends to be the fastest accumulation of plastic waste. Much of this generated plastic waste is buried in landfills, floating across the ocean, clogging drainage, dispersed in the air as microplastics, littering roads, or in other parts of the environment.

While 7 African countries: Algeria, Nigeria, Ecuador, Gabon, Angola, Equatorial Guinea, and Conge, are part of the organization of the petroleum exporting countries (OPEC, 2022), 38 out of 53 African countries are net oil importers (African Development Bank, 2009). Many African countries have limited access to modern energy infrastructure. Ethiopia, for example, derives 88% of its energy from biomass (Yalew, 2022). These are mainly woody biomass and agricultural waste. Ethiopia exploits 50% of its exploitable woody biomass resource and 30% of its agricultural waste resource for energy use but exploits less than 10% of its exploitable hydropower resource and less than 1% of its exploitable solar, wind, and geothermal energy. Only around 3% of the energy consumption in Ethiopia is electricity, 95% of which is generated from hydropower. It is estimated that 56% of Ethiopians do not have access to electricity with Ethiopia being ranked as having the lowest energy consumption per capita (Yalew, 2022). Around 98.9 million Ethiopians cook using biomass energy as of 2018 (Oyewo et al., 2021).

In Uganda, 0.73 kWh of energy is consumed per capita (Amulen et al., 2022). Globally, energy demand is on the rise. The projected increase for 2021 was 4.6%. This increased demand is attributed to the increasing demand for energy by emerging markets and developing economies (IEA, 2021). Coal, gas, and fossil oil are all showing increasing demand. Africa is not exempt from this increase in demand for energy as much of the continent is emerging markets and developing economies. The generation of electricity from renewable sources is projected to grow in the coming years as the world moves towards more sustainable energy generation. While chemical recycling and energy recovery from plastics and polymer waste are not renewable, it is, however, the more sustainable approach to better utilize non-renewable resources used in the production of these plastics and polymers.

Plastics and polymer waste contain considerable amounts of energy and can be converted to other valuable products such as fertilizers and construction materials as discussed within this chapter. The chemical recycling and energy recovery from this category of waste can help meet the energy need of African countries and also mitigate against the adverse impact of discarding these materials in the environment. Developing efficient waste management systems and chemical recycling and energy recovery technology advancement is needed to meet this potential.

A study of waste composition in Africa showed that on average, waste in African countries comprises around 57% (Hoornweg & Bhada-Tata, 2012) of the organic components and is generally wetter. This makes it not well suited for energy recovery (ACCP, 2019). Composting and burying organic waste separated from other inorganic waste might be a suitable low-cost and low-tech option for recovering value in the form of soil nutrients to feed farms and gardens.

Conclusion

Chemical recycling and energy recovery from plastic waste fall under the 7th sustainable development goal set out by the United Nations which addresses meeting the need for affordable, reliable, and efficient energy supply for all towards alleviation of poverty (UN, 2021). Challenges to large-scale implementation of chemical recycling and energy recovery from waste plastic and municipal solid waste, in general, include the variation in waste stream composition, possibilities of the presence of hazardous and electrical components, and availability of adequate technology. These can be addressed by setting policies and regulations to prohibit the dumping of hazardous materials and E-waste in landfills, private–public partnerships, and measures such as charging gate fees per tonne of the waste deposited at landfills to generate income to develop chemical recycling and waste recovery facilities. While there is a global need for advancing the chemical recycling and energy recovery technology and application, understanding the current status and need case by region is important to develop strategies and innovations that are suited to each region. Hence, the focus of this chapter is on the African continent.

References

ACCP. (2019). *Africa solid waste management data book 2019*. Ministry of the Environment of Japan, UNEP, UN-Habitat.

African Development Bank, 2009African Development Bank. (2009). *Oil and gas in Africa* (pp. 1–2). Oxford University Press. ISBN: 978-0-19-956578-8.

Ahaneku, I. E., & Njemanze, C. F. (2015). Material flow analysis of abattoir solid waste management system in Minna, Nigeria. *Journal of Solid Waste Technology Management, 41*(2), 165–172.

Amulen, J., Kasedde, H., Serugunda, J., & Lwanyaga, J. D. (2022). The potential of energy recovery from municipal solid waste in Kampala City, Uganda by incineration. *Energy Conservation and Management: X* (in press).

Audu, I. G., Barde, A., Yila, O. M., Onwualu, P. A., & Lawal, B. M. (2020). Exploring biogas and biofertilizer production from abattoir wastes in Nigeria using a multi-criteria assessment approach. *Recycling, 5*(3), 1–24.

Bakalar, T., Pavolova, H., Hajduova, Z., Lacko, R., & Kysela, K. (2021). Metal recovery from municipal solid waste incineration fly ash as a tool of a circular economy. *Journal of Cleaner Production, 302*, 126977.

Desta, M., Lee, T., & Wu, H. (2022). Life cycle energy consumption and environmental assessment for utilizing biofuels in the development of a sustainable transportation system in Ethiopia. *Energy Conservation and Management: X, 13*, 100144.

Ekwall, T., Gottfridsson, M., Nellstrom, M., Nilsson, J., Rydberg, M., & Rydberg, T. (2021). Modelling incineration for more accurate comparison to recycling in PEF and LCA. *Waste Management, 136*, 153–161.

Geyer, R., Jambeck, J., & Law, K. (2017). Production, use, and the fate of all plastics ever made. *Science Advances, 3*(7), e1700782. https://doi.org/10.1126/sciadv.1700782

Ginder, R. S., & Ozcan, S. (2019). Recycling of commercial E-glass reinforced thermoset composites via two temperature step pyrolysis to improve recovered fiber tensile strength and failure strain. *Recycling, 4*(24). https://doi.org/10.3390/recycling4020024

Hoornweg, D., & Bhada-Tata, P. (2012). *What a waste: A global review of solid waste management.* World Bank. Urban Development Series Knowledge Paper. 15.

IEA International Energy Report. (2021). Global energy review. Flagship report April 2021. Retrieved 17 February, 2022, from https://www.iea.org/reports/global-energy-review-2021

Jambeck, J., Hardesty, B. D., Brooks, A. L., Friend, T., Teleki, K., Fabres, J., Beaudoin, Y., Bamba, A., Francis, J., Ribbink, A. J., Baleta, T., Bouwman, H., Knox, J., & Wilcox, C. (2018). Challenges and emerging solutions to the land-based plastic waste issue in Africa. *Marine Policy, 96*, 256–263.

Kefalew, T., & Lami, M. (2021). Biogas and bi-fertilizer production potential of abattoir waste: Implication in sustainable waste management in Shashemene City, Ethiopia. *Heliyon, 7*, e08293.

Lambreton, L., & Andrady, A. (2019). Future scenarios of global plastic waste generation and disposal. *Palgrave Communications, 5*(6). https://doi.org/10.1057/s41599-018-0212-7

Lopez, G., Artetxe, M., Amutio, M., Alvarez, J., Bilbao, J., & Olaza, M. (2018). Recent advances in the gasification of waste plastics. A critical overview. *Renewable and Sustainable Energy Reviews, 82*(1), 576–596.

Moretti, C., Junginger, M., & Shen, L. (2020). Environmental life cycle assessment of polypropylene made from used cooking oil. *Resources Conservation and Recycling, 157*, 104750.

Nabgan, W., Nabgan, B., Abdullah, T. A. T., Alqaraghuli, H., Ngadi, N., Jalil, A. A., Othman, M. B., & IbrahimMA, S. T. (2020). Ni–Pt/Al nano-sized catalyst supported on TNPs for hydrogen and valuable fuel production from the steam reforming of plastic waste dissolved in phenol. *Internation Journal of Hydrogen Energy, 45*(43), 22817–22832.

Oliveux, G., Dandy, L. O., & Leeke, G. A. (2015). Current status of recycling fiber-reinforced polymers. *Progress in Material Science, 72*, 61–99.

OPEC Member Countries. Retrieved 11 February, 2022, from https://www.opec.org/opec_web/en/about_us/25.htm

Oyewo, A. S., Solomon, A. A., Bogdanov, D., Aghahosseini, A., Mensah, T. N. O., Ram, M., & Breyer, C. (2021). Just transition towards defossilised energy systems for developing economies: A case study of Ethiopia. *Renewable Energy, 176*, 346–365.

Pandey, U., Stormyr, J. A., Hassani, A., Jaiswal, R., Haugen, H. H., & Moldestad, B. M. E. (2020). Pyrolysis of plastic waste to environmentally friendly products. *Energy Production and Management in the 21st Century IV, 246*, 61–74.

Pereira, E. G., & Martins, M. A. (2017). Gasification technologies in encyclopaedia of sustainable technologies. *Reference Module in Earth Systems and Environmental Sciences*, 315–325.

Qin, L., Xu, Z., Zhao, B., Zou, C., Chen, W., & Han, J. (2022). Kinetic study on high-temperature gasification of medical plastic waste coupled with hydrogen-rich syngas production catalyzed by steel converter ash. *Journal of the Energy Institute, 102*, 14–21.

Rehan, M., Nizami, A. S., Asam, Z. Z., Ouda, O. K. M., Gardy, J., Raza, G., Naqvi, M., & Ismail, I. M. (2017). Waste to energy: A case study of Madinah city. *Energy Procedia, 142*, 688–693.

Seabea, D., Ruengrit, P., Arpornwichanop, A., & Patcharavorachot, Y. (2019). Gasification of plastic waste for synthesis gas production. *Energy Reports, 6*, 202–207.

Shuma, R., & Madyira, M. D. (2019). Emissions comparison of loose biomass briquettes with cow dung and cactus binders. *Procedia Manufacturing, 35*, 130–136.

Simegn, A., Abebe, S., & Worku, A. (2021). Characterization and optimization of incinerated municipal solid waste fly ash as a cement substitute material in concrete at Reppie waste to energy plant in Ethiopia, East Africa. *Advances in Environmental Studies, 5*(1), 382–393.

Sindibu, T., Solomon, S. S., & Ermias, D. (2018). Biogas and bio-fertilizer production potential of abattoir waste as means of sustainable waste management option in Hawassa City, Southern Ethiopia. *Journal of Applied Science and Environment Management, 22*(4), 553.

Statista. (2021b). *Global polyethylene and polypropylene capacity 2017–2023.* Retrieved 10 March, 2022, from https://www.statista.com/statistics/1118115/global-polyethylene-polypropylene-capacity/#:~:text=In%202017%2C%20the%20total%20production%20capacity%20of%20polyethylene,to%20some%20230%20million%20metric%20tons%20in%202023

Statista. (2021a). *Global polyethylene demand and capacity 2015–2022.* Retrieved 10 March, 2022, from https://www.statista.com/statistics/1246675/polyethylene-demand-capacity-forecast-worldwide/

Torres, A., de Marco, I., Caballero, B. M., Laresgoiti, M. F., Legarreta, J. A., Cabrero, M. A., Gonzalez, A., Chomon, M. J., & Gondra, K. (2000). Recycling by pyrolysis of thermoset composites: Characteristics of the liquid and gaseous fuels obtained. *Fuel, 79*(8), 897–902.

UNEP. (2013). *The Japanese industrial waste experience: Lessons for rapidly industrializing countries* (pp. 94–106).

UNEP. (2018). Africa waste management outlook. *Summary for decision-makers.*

UNEP. (2018). *Single-use plastics: A roadmap for sustainability.*

United Nations Department of Economic and Social Affairs Statistics Division. *The sustainable development goals report 2021.* Retrieved 12 March, 2022, from https://unstats.un.org/sdgs

Wu, C., & Williams, P. T. (2009). Hydrogen production by steam gasification of polypropylene with various nickel catalysts. *Applied Catalysis B, 87,* 152–161.

Yalew, A. W. (2022). The Ethiopian energy sector and its implications for the SDGs and modeling. *Renewable and Sustainable Energy Transition, 2,* 100018.

Yoshimura, M., & Byrappa, K. (2008). Hydrothermal processing of materials: Past present and future. *Journal of Materials Science, 43,* 2085–2103.

Zhao, D., Wang, X., Miller, J. B., & Huber, G. W. (2020). The chemistry and kinetics of polyethylene pyrolysis: A process to produce fuels and chemicals. *Chemsuschem, 13*(7), 1764–1774.

Zong, Y., Li, F., & Liu, Z. (2019). Extraction of alumina from high-alumina coal ash using an alkaline hydrothermal method. *SN Applied Science, 1,* 783.

Chapter 10
Plastic and Polymer Waste Management Systems in Africa

Abstract A sustainable plastics and polymer industry also includes the sustainable management of the waste generated from the industry. Management of waste includes collecting the waste, recovery of value from the waste, disposing of the waste, as well as reducing the generation of the waste. Management of waste requires the input of resources, time, labor, and skill, either during the process of manufacturing products or the waste generated at the end of the usage life of the product. To this end, this chapter is dedicated to reviewing the waste management systems of plastic and polymers with a specific focus on the African region. The waste category of interest here is a municipal solid waste of which plastics and polymers are a large proportion. The chapter begins by looking at the existing data and statistics of global waste management and then specifically at waste management statistics and the current status in the African region. The sections that follow explore the formal and informal waste management systems in selected countries in Africa. A section is dedicated to observations of informal waste management of some polymer-based waste materials within Lagos Nigeria.

Keywords Waste management · Recycling · Waste composition · Landfill · Plastic sorting · Engineered landfill

Global Waste Management

As more of the world gets urbanized, one of the greatest markers of urbanization; waste generation, increases as a result. Currently, the rate of increase in waste generation rate is higher than the rate of urbanization. The last World Bank report on global waste generation (Hoornweg & Bhada-Tata, 2012), estimated the municipal solid waste generated globally to be around 1.3 billion tonnes every year. By 2025, this is projected to increase by another 0.9 billion to reach 2.2 billion tonnes of solid waste generated in the world every year. It is estimated that the average person generates about 0.64 kg of municipal solid waste a day. With this comes an increasing cost of managing this waste and the increasing impact of the waste on health and the environment.

© The Author(s), under exclusive license to Springer Nature Singapore Pte Ltd. 2022   147
O. Olatunji, *Plastic and Polymer Industry by Region*,
https://doi.org/10.1007/978-981-19-5231-9_10

China is currently the highest waste generator in the world with the US being the second highest waste generator. Rural areas generally generate less waste than urban areas. This is attributed to the fact that rural areas tend to depend less on packaged goods which contribute significantly to waste generation. The rural areas are also more likely to reuse, repurpose, and recycle items. The rural–urban waste generation pattern becomes more complex where increased consumption of packaged goods is not necessarily a result of increased disposable income, rather it is a result of more people living in cities and having no alternatives to packaged goods. An example is sachet water in parts of Nigeria which serves as the only source of clean drinking water for many in underserved urban areas.

Already it costs the world about 205.4 billion USD to manage municipal solid waste every year and rising. The ability of a country to manage its municipal solid waste is reflected in its ability to manage other aspects of society such as education, health, and transportation. Often these factors are linked. For example, a society with a high level of education and awareness is more likely to take individual responsibility to separate and properly dispose of household and commercial waste. Such a society is likely to have good road networks that allow waste collection trucks to access all communities and this leads to a cleaner environment and fewer health hazards from waste pollution.

Waste has become a global issue. Household waste from one location can find its way across the oceans to end up at a beach on another continent. Products created in one country often end up as waste in another country where it has been imported and discarded after use. Furthermore, the greenhouse gas emitted from waste becomes a global issue once the greenhouse gases get released into the atmosphere. Essentially, the world shares the same water and breathes the same air.

The global challenge in terms of municipal solid waste management is to find an effective way to get rid of the waste at minimal cost while avoiding any adverse impacts to the environment and health such as greenhouse gas emissions and release of toxins into the environment. When possible, utilize the waste to generate energy and/or materials. To do so, it is important to understand the systems that exist in different parts of the world. Hence, the aim of this chapter is to explore waste management systems as it exists within a continental region, Africa.

The six main waste management practices (Fig. 10.1) are source reduction, waste collection, recycling, composting, incineration, and landfilling or dumping. The extent to which each of these is achieved and the mode of operation of ovaries for different regions also varies within the same region. For example, in the OECD regions and countries classified as high-income countries, as high as 98% of waste generated gets collected while in countries classified as low income, around 41% of waste generated gets collected (Hoornweg & Bhada-tata, 2012). This can be explained by high-income countries having more funds to dedicate to waste management. Furthermore, within the same country, waste management in the urban areas varies from that in the rural areas.

Municipal solid wastes include waste from households, industries, commercial activities, institutions, constructions, and demolitions and waste generated from

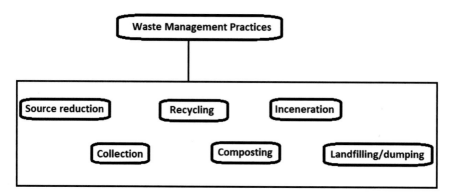

Fig. 10.1 Waste management practices

municipal services such as street cleaning, waste from beaches, and community beautification.

Composition of Waste Generated in Africa

Estimates for different regions report that 57% of the waste generated in Africa is organic materials, 13% of the waste is plastic, 9% is paper, metals make up 4%, and glass also makes up 4%. Other materials which do not fall into these categories make up 13% of the waste generated in the African region as in 2012 data (Hoornweg & Bhada-Tata, 2012). This is summarized in Fig. 10.2. The organic portion of this waste comprises polymer-based materials such as polymers in food, cosmetics, and plant materials. Paper also comprises polymers in the form of cellulose and polymers like starch used as additives. Materials like rubber and fabrics fall into the "other" categories. It can therefore be said that at least over 80% of waste generated in the African continent comprises plastics and polymers in various forms.

The distribution of waste composition in Africa is similar to the global average distribution where a large proportion of the waste generated across the world is organic (46%), and metal and glass make up the least proportion (4% and 5%, respectively), plastics make up 10% and paper makes up 17%. Other materials make up 18% of the global waste distribution. The main difference between waste composition in Africa and in the OECD countries is that a higher composition of waste in Africa is organic, while in the OECD countries, organic waste composition of total waste generated is much lower at around 27%. The OECD countries also have a larger proportion of paper (32%) in the waste compared to African countries.

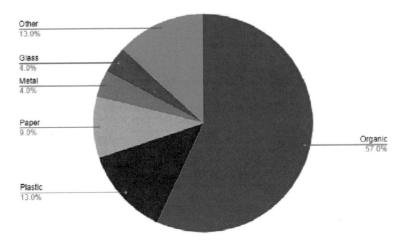

Fig. 10.2 Waste composition estimated for the African region

An Overview of Waste Management Across Africa

The waste management issue existed as early as ancient Rome where dumping garbage and excrement on the outskirts of the city resulted in stench and communicable diseases. Medieval London also dealt with epidemics that spread as a result of improper waste management and deteriorated public sanitation. Japan began implementing the separation of waste around the 1970s, however, not until the 1990s did it pass legislation on controlling emissions and recycling. (ACCP, 2019). Across the world, countries face different forms of waste management challenges and continue to pass various laws relating to extended producer responsibilities.

Each part of the world has its own specific challenge when it comes to waste management. Composition of waste, attitude to waste, and technical capacity to handle waste among other issues determine the nature of waste management in any given region, country, or city. Prior to urbanization, rural areas of Africa traditionally generated mostly organic waste. For such waste, the natural environment was capable of disposing of the organic wastes with a minimal rate of accumulation. Waste was largely food residues, excrement, and by-products of agricultural produce. These were reused as animal feed, textiles, or soil nutrients.

It is reported that in 2012 the amount of waste generated within the African continents was around 125 million tonnes and 90% of this ends up in uncontrolled dumpsites and landfills (UNEP, 2018). In sub-Saharan Africa, it is reported that around 70% of the waste generated ends up in open dump sites (ACCP, 2019). There have been reports in recent years of these dumpsites resulting in accidents such as fires, collapses, and landslides that resulted in fatal casualties. Between 2017 and 2018, dumpsite collapse that resulted in between 9 and over 100 victims have been reported in Guinea, Ethiopia, and Mozambique (ACCP, 2019).

Analysis shows that the volume of waste generated annually within the African continent by 2050 is expected to increase from 174 million tonnes to an estimated 516 million tonnes (Tomita et al., 2020). An estimated 13% of this waste is plastic waste and 57% is found to be an organic waste, part of which includes polymeric materials like cellulosic materials and proteins and polysaccharides in food waste. With the current average waste collection rate of 55% and 90% of the waste generated is disposed of at dumpsites and landfills, it will be difficult to keep up with this projected rise in waste accumulation without significant improvement in waste management systems. Sub-Saharan Africa has 19 of the world's biggest dumpsites.

Landfilling and dumps are the most commonly applied waste management practices in the African region. Other waste management practices are less applied. In OECD countries, landfilling as well as thermal treatments are most used, however, other waste management practices such as composting and recycling are substantially applied (Hoornweg & Bhada-Tata, 2012).

When considering the total annual waste generation across the world, Africa and South Asia generate the least amount of waste. Per capita, Africans generate on average 0.65 kg of waste per day. OECD countries generate on average 2.2 kg of waste per day, while those in the South Asia Regions generate on average 0.45 kg per capita per day. Africa has an estimated urban population of 260 million people generating a total of 169,119 tonnes of municipal solid waste daily. OECD countries have an urban population of 729 million generating 1,566,286 tonnes of municipal solid waste daily, while the South Asian Region has an urban population of 426 million generating 192,410 tonnes of municipal solid waste daily (Hoornweg & Bhada-Tata, 2012). These values vary slightly for different data sources. For example, ACCP reports an average of 0.6 kg per capita daily waste generation based on surveys from 23 cities (ACCP, 2019). Therefore each person in OECD countries generates more waste on a daily basis than each individual in African countries on average. Also as a region, Africa generates less waste when compared to all the other regions.

Management of plastic and polymer waste results in greenhouse gas emissions. The decomposition of biodegradable polymers results in the production of greenhouse gases, the vehicles used in the collection of this waste emit greenhouse gases and often consume fossil fuels the different methods of processing and conversion of the waste may also result in the further emission of greenhouse gases. The goal is therefore to effectively manage the waste generated by society such that the greenhouse gas and fossil fuel consumed in the management of the waste is minimal. Furthermore, the waste should have no adverse impact on the environment but rather be useful and beneficial as energy or material.

According to a world bank document on a global review of solid waste management (Hoornweg & Bhada-Tata, 2012), solid waste management activities can be classified into; source reduction, collection, recycling, composting, incineration, landfilling, or dumping. When evaluating the cost of any particular waste management practice, the financial cost, as well as the cost to the environment, should be assessed in both the short and long terms. A low-cost waste management practice in terms of monetary cost might have a higher cost in terms of impact on the environment.

While in regions like Europe and countries like Japan, sorting of household waste has been enforced for the past three decades, most sub-Saharan countries do not have any enforcement on household waste sorting. Much of the recycling done in the African region is intensified by financial profit from recycling. In North America, for example, the recycling rate is around 90%. The average recycling rate for sub-Saharan Africa is 44% (ACCP, 2019).

A study on the impact of environmental awareness and behavior impact on health in Kenya found that around 80% of diseases in developing countries, which includes countries in the African continent, are associated with improper management of waste (Selin, 2013). For example, waste dump sites act as a breeding ground for disease-carrying rodents and malaria-causing mosquitoes or the contamination of soil and drinking water by seepage from dumpsites.

Accessibilities of some areas are a challenge in some cities in some African countries. For example, the ACCP reports that in Congo Brazzaville, 65% of the city areas cannot be easily accessed by vehicles. Maintenance of equipment and facilities for waste management is also a challenge. Some cities with more effective waste management systems receive support from international organizations (ACCP, 2019). This includes grants, procurement of vehicles and equipment, and technical capacity building.

Some cities have transfer points where waste collection by wheelbarrows and donkey carts in smaller quantities are accumulated in a contained space to be collected by larger vehicles and taken over longer distances to disposal points. This reduces the cost of transporting waste from points of generation. It is a particularly useful system where roads are inaccessible by vehicle. The size of the transfer points varies from small scale to large scale.

Sanitary landfills exist in cities like Maputo in Mozambique, Kinshasa in DR Congo, Ouagadougou in Burkina Faso, Monrovia in Liberia, Lusaka, and Bulawayo in Zimbabwe, Yaounde in Cameroon, Djibouti in Djibouti, Tema in Ghana, Alexandria in Egypt, Windhoek in Namibia, and Kweneng in Botswana. Controlled dumpsites are used in Blantyre in Malawi, Juba in South Sudan, Antananarivo in Madagascar, Harare in Zimbabwe, and Abuja in Nigeria. Others like Khartoum in Sudan, Kaduna in Nigeria, Nairobi and Kiambu in Kenya, Brazzaville in Congo, Maseru in Lesotho, Conakry in Guinea, Addis Ababa in Ethiopia, and Niamey in Niger operate open dumpsites. Whether sanitary landfill, controlled or open dumpsite systems are used, the operation of the waste management system varies for each case. A landfill designed as a sanitary landfill will only serve as one if it is operated effectively with a good liner, leachate control and treatment, and compaction and soil cover carried out efficiently.

Observations of Open Plastic Waste Dumping

Based on observations, it can be said that openly discarded waste and heaps of waste consisting to a large extent of plastics, can be seen across many parts of Lagos and other towns and cities in Nigeria. Figures 10.3, 10.4, 10.5, 10.6, 10.7, 10.8, 10.9 and 10.10 shows images of plastic waste disposed of along roads, blocking drainages, and other places.

Formal Waste Management Systems in Some Countries in Africa

This section looks at waste management authorities in selected countries across Africa. The activities of the organizations are discussed and recent developments relating to the organizations are also discussed. These include public, private, and public–private partnership organizations. While there are several others not discussed here, this is expected to provide the reader with a broad idea of different waste management organizations across the region.

The key challenges of municipal solid waste management in African countries include limited coverage of the available waste management services, insufficient or inefficient recycling facilities, improper landfill disposal, and improper handling of hazardous and medical wastes. Part of the factors related to these issues includes limited road access in some parts which, for example, prevents the garbage truck from accessing and unplanned developments which make it difficult to optimally locate public waste management facilities such as recycling plants and landfills.

ACCP reports that in a survey involving 25 ACCP member countries, 64% of the countries' ministry of the environment are responsible for solid waste management. These ministries had different bodies responsible for controlling solid waste management. In other countries, solid waste management was the responsibility of local government, department of public works, and ministries for utilities or other authorities.

Solid waste management legislation in many African countries is still in its infancy. Laws clearly defining responsibilities regarding recycling, hazardous waste disposal, and energy recovery from waste, for example, are not established in many African countries (ACCP, 2019). Recent years have seen many African countries putting in place laws to mitigate plastic pollution problems. Countries like Rwanda have successfully enforced these laws. Improving financing and technical capacity in waste management will aid the effective establishment and enforcement of waste management laws.

In Lagos, for example, there are some visible efforts by government and private partners to provide public waste management services and facilities. For example, Fig. 10.11 shows some waste management officers at work and the waste bin

(a)

(b)

Fig. 10.3 **a** plastic dump by a bush within a residential area in Lagos, Nigeria, January 2022 **b** A close-up image of a goat (in black circle) openly grazing amid the waste and likely to ingest plastic

Fig. 10.4 Plastic waste observed along the NNPC road, Lagos, Nigeria. January 2022

provided at a public park. It is also evident that the public need to cooperate and take responsibility for their waste by making proper use of the bins provided.

Fig. 10.5 Plastic waste observed along the Lagos-Ibadan railway line. March 2022

Fig. 10.6 Plastic waste (Circled in black on the left-hand image) pictured at the ancient Olumo Rock, a popular tourist attraction in Abeokuta in Ogun State, Nigeria January 2022

Fig. 10.7 Individuals were observed cleaning out plastic from road drainage in front of shops along the Ikotun Egbe Road, Lagos, Nigeria April 2022

Rwanda Environmental Management Authority (REMA)

The Ministry of Local government is responsible for waste management in Rwanda implemented by the government-owned Water and Sanitation Corporation (WASAC) Ltd. Private companies participate in waste management mainly as waste collection services. Apart from Kigali which has its own waste management, the districts manage waste management in the other four provinces; included in this waste are plastic wastes and other polymer wastes such as textiles, biomass, and rubber.

REMA exists under the Ministry of Environment. It was established in 2013. It is dedicated to national environmental protection, conservation promotion, and management. Its activities include providing advice to the government in areas relating to the environment and climate change. REMA has been included in this section as the management of plastics and other polymer wastes have in recent times been a high priority globally. REMA is also committed to assisting the government in the area of sustainable use of natural resources. In carrying out its role its activities include disseminating information to the general public on the environment,

Fig. 10.8 Plastic waste blocking drainage on a street at Okota, Lagos, Nigeria. May 2022

carrying out environmental audits and assessments, and providing technical support to private and public entities in the area of environment conservation and sustainable management of natural resources (www.rema.gov.rw/about).

Beyond activities directly relating to the environment and sustainable resource management, REMA also collaborates with other government agencies on various

Fig. 10.9 Plastic bottles observed floating on the Lagos lagoon. August 2019

projects which allow it to have a wider impact beyond its mandate. For example, REMA in collaboration with the ministry of infrastructure, Local Administrative Entities Development Agency, and the City of Kigali implemented the world bank-funded Rwanda Urban Development Projects with the second phase of the project beginning in March 2022. Impacts of these projects extended to improved safety through better urban planning that creates street lights and safe walking paths and roads, improved livelihoods through the restoration of natural environments to create Eco Parks, cleaning up water bodies, and much more (https://www.rema. gov.rw/info/details?tx_news_pi1%5Baction%5D=detail&tx_news_pi1%5Bcontrol ler%5D=News&tx_news_pi1%5Bnews%5D=501&cHash=c5fb0315663dc41d83 8dd65658e051b6).

A study of the waste management systems in the City of Kigali (Kabera & Nishimwe, 2019) reported that only one dump site existed to serve the city as of the time of the study. Kigali the capital city of Rwanda with a population estimated at 1.5 million accounts for a tenth of the country's population. The city generates around 638 tonnes of waste daily which is around 0.41 kg per capita per day according to estimates based on 2012 data.

The collection and transportation aspect of waste management in Kigali is handled by private companies. These companies charge households according to locations and income class. The waste collected is then transported to the Nduba disposal site in the Gasabo district of Kigali. This dumpsite is run by the authorities of the City

Fig. 10.10 Plastic dumped around and overflowing waste bin at a motor park in Iyana Ipaja, Lagos June 2022

of Kigali. The study estimates the rate of recycling to be 10%, however, official figures for the recycling rate were previously 2%. Much of the recycling is also done by private companies. In Kigali households, commercial establishments are known to have cultivated the practice of separating their waste. This makes it easier for recyclers to obtain recyclables and also reduces the burden on the dumpsite. Some of the recyclables are also exported to the neighboring countries; Uganda and Tanzania to be recycled. These can be attributed to these countries having a more active plastic recycling industry than Rwanda where laws exist against the manufacturing of single-use plastics. The country is also more focused on the service sector than manufacturing.

It is suggested that the waste management system in Rwanda can be improved through increased investments from the private sector incentivized and augmented by the government budget allocation. The government should also strategically create a positive relationship between the private waste companies and the communities from which they collect such that the communities are encouraged to pay for and support these waste collection services. Rwanda has an inclusive waste management system

Fig. 10.11 **a** Workers picking plastic waste at a park in Lagos, Nigeria, **b** Public bins provided at the park, **c** Plastic packaging improperly discarded on the park grounds by visitors to the park despite the public bins provided. April 2022

(c)

Fig. 10.11 (continued)

where communities are provided with opportunities to voice concerns relating to waste management to authorities. It also has legislation and regulations on solid waste management. To improve its waste management system, there need to be facilities such as waste treatment and leach prevention systems and a weighbridge at the dumpsite to improve the collection of data on waste disposal. Table 10.1 gives a list of some waste management authorities in some cities and countries on the African continent (Yoada et al., 2014).

Municipal Solid Waste Management in Senegal

In Senegal the management of municipal solid waste is the responsibility of local authorities at the state level. In Senegal for example, an estimated 966.155 tonnes

Table 10.1 Some waste management authorities in some countries in Africa

Waste Management Authority/Agency	Location	Private/Public
LAWMA	Lagos, Nigeria	Public
KCCA PPP Kampala capital city authority Private Public Partnership	Kampala, Uganda	PPP
Division Régionale de l'Environnement et des Établissements Classés Service Regional de l'hygiene	Senegal	Public
MMDA (Metropolitan, Municipal, and District Assemblies under the Ministry of Local Government and Rural Development	Ghana	Public
REMA (Rwanda Environment Management Authority	Rwanda	Public
Addis Ababa City Solid Waste Management Agency	Addis Ababa, Ethiopia	Public

of municipal solid waste is generated annually and around 237.25 kg is generated annually per capita. In the city of Dakar, 172.11 kg of waste is generated annually per capita with a recycling rate of around 7.44%. It is reported that 80% of the waste gets collected. Landfilling is an open dump is the only known formal way waste collected is managed in Dakar. The Mbeubeuss waste dump serves as the only waste dump for the city (Beri, 2018). Private processors informally collect plastics, cardboard, metal, glass, and other recyclables. Enforcement of an existing tax system is projected to generate around 20 billion CFA francs that can be used for the development of waste management infrastructure in the city. However, only around 10% of this tax is currently being recovered by the state.

Senegal is currently undergoing rapid urbanization and population growth which has resulted in changing consumption patterns (German Federal Ministry for Economic Cooperation and Development, 2021). A large proportion of Senegal's food supply, around 70% is imported with fishing and tourism being the main source of GDP. This means that maintaining a clean environment is important for these two economic activities. Plastic pollution in the waters will threaten fishing, while an unclean environment will put off tourists. In previous years, SOADIP, SIAS, and other companies had been the main players in waste management in Dakar, the capital city of Senegal. However, these companies seized operations due to ineffective management and finance. Waste management in the city has since been carried out by Veolia, A French company. The main activity of the company is the collection of waste from the municipalities and dumping at the Mbeubeuss waste dump.

Reports suggest that households contribute around 80% of the waste produced in Senegal and the urban areas generate more waste than the rural areas. Forty-four percent of the waste generated in the city of Dakar is organic waste consisting of garden waste, fisheries waste, poultry, groundnut shells, industrial waste, fertilizers, and others. Plastics make up 18% of the waste generated in the city of Dakar and most of it comes from the plastic processing industries.

Dakar operates on an integrative waste management system where the public and private sectors are jointly involved in waste management. The upstream waste management which generally comprises collection and dumping is handled by the local authorities and the main private company in waste management in Dakar, Veolia. The recycling of materials like plastics is mainly done by local recyclers on a small scale. This generally involves scavengers collecting waste from the dumpsite. Organic wastes like rice are collected and sold as animal feed.

In Thies, the third most populous region in Senegal, recent years have seen increased urbanization with more industries, an increase in tourism and hospitality businesses, and increased trade infrastructures. This has led to an increase in waste generation and thus more need for waste management. Like other parts of Senegal, most of this waste ends up at the dumpsite. In Thies, the main dumpsite is a space that formerly served as a quarry (IPEP, 2006). Being a former quarry the land space has a pre-existing hollow into which waste is dumped. It is estimated that 110,000 cubic meters of waste are generated from Thies annually. This includes plastic and polymer wastes like plastic bags, rubber, paper, and food wastes. The local authorities are responsible for waste management in the region.

New developments in waste management in Senegal include the development of engineered landfills in Saint Louis, Thies, and Touba. These landfills are better contained with a black plastic barrier to prevent seepage from dumped waste getting into the soil. There are also ongoing projects towards implementing the three Rs of sustainability in different regions through the regulatory systems as well as raising awareness (IPEP, 2006).

Management of Plastic Waste in Kenya

There have been reports of endangered species of turtles on the beach of Watamu in Kenya getting harmed from ingesting waste plastics. One study on the attitude to recycling by households in Watamu, Kenya found that of the households studied, 0.7% recycled the plastic waste they generated (Gwada et al., 2019). The plastics discarded were mainly HDPE, LDPE, PET, and PP. 61.4% of the plastic waste generated by households ended up in dumpsites; 12.9% were burnt and 6.4% were discarded in other ways. There is a high level of reuse of plastics by many households in this region as this study suggests. Although recycling activities in Kenya are mostly done by small-scale entrepreneurs, around 93.6% of households reuse plastic packaging for other purposes such as food, water, and oil storage.

In Kenya, Nairobi and Kiambu are two of the states that are part of the ACCP. An assessment of waste management in ACCP member states (ACCP, 2019) reports that Kiambu has an open dump system where compaction is carried out, while in Nairobi, the capital city, an open dump system persists with no compaction.

Solid Waste Management in Kampala

Taking the case of Kampala, the capital city of Uganda, a country in East Africa, waste management is the responsibility of the Kampala Capital City Authority KCCA. It is responsible for the collection, transport, treatment, and disposal of municipal solid waste generated in the greater Kampala Metropolitan. In 2010, KCCA sponsored the Kampala waste management project under the KCCA Act of 2010 (KCCA, 2020). The project went into operation in 2011 with the mandate to provide waste management and environmental conservation services to residents and businesses in the city of Kampala. The project is a public–private partnership jointly funded by the private sector and the Ugandan government centered around the development of the Ddundu landfill.

In Kampala, there is the Kiteezi landfill which the KCCA sets to close down (Aryampa et al., 2020) and replace with a sanitary engineered landfill to be located at Ddundu. Kampala has a population of around 2 million residents with tourists and other visitors making up a transient population of 5 million. As the capital city, Kampala is the center of the rapid urbanization and industrialization which Uganda is undergoing. As are characteristics of increased urban lifestyle, waste generation is also on the rise. The KCCA set out the Kampala waste management project to promote sustainable environmental activities and mitigate climate change in addition to developing an effective solid waste management system for the city.

Data from KCCA shows that waste collection in Kampala has increased in recent years with an annual compounded growth rate of 7%. This is attributed to the private company's increased involvement in waste collection. As of 2012, an estimated 344,149 tonnes of waste was collected in Kampala. This rose to 481,082 by 2017. KCCA collected 67% of the waste in 2012 and the remaining 33% of waste collected in that year was by the private sector. However, by 2017, the private sector collected 45% of the waste collected in Kampala for that year, while KCCA collected 55%. It is estimated that waste collection in Kampala has an efficiency of 60%. Kiteezi has been the only authorized landfill in Kampala since its construction in 1995. It is, however, not a sanitary engineered landfill without a liner or drainage channels to prevent the seepage of contaminant waste into soil and nearby water bodies. The steepness of the landfill also poses the risk of landslides.

Analysis shows that Kampala's municipal solid waste in the Kiteezi landfill comprises 43% biodegradable waste, 6% plastics, 1% textile, 2% paper/ cardboard, 3% animal waste, 3% sanitary waste, 42% mixed fines, and 1% other wastes that do not fall into any of these categories (KCCA, 2020). The waste comprises a relatively low calorific value and low energy value of around 6.12 MJ/kg such that the prospect of waste to energy plant is not high.

Plastic Waste Management in Nigeria

Nigeria is the most populous nation in Africa with a population near 200 million. Its economy is also regarded as one of the largest in Africa with a 375.75 billion USD GDP as of 2017. Solid waste management in the country falls under the responsibility of the Department of Pollution Control, Solid Waste Management, and Technology which exists under the Ministry of Environment. However, other bodies such as the Ministry of Health, the Ministry of Agriculture and Rural Development are also tasked with the responsibility of medical and agricultural solid waste management, respectively. Solid waste management in the Federal Capital Territory is the responsibility of the Abuja Environmental Protection Board. Currently, no basic law exists on municipal solid waste management and there are no sanitary landfills in the country (ACCP, 2019).

The media has also been used in promoting awareness of the proper management of waste within the country. For example, radio station programs have proven useful in raising awareness and positive attitude of citizens in Yobe state in Nigeria on proper ways to handle and dispose of solid waste (Bappayo & Maidunoma, 2018). Despite the limited waste management infrastructure, studies suggest that recycling is higher in lower income households compared to higher income households, for example, according to studies carried out in the Kaduna state of northwest Nigeria (Abd'Razack et al., 2016). Lower income households are more likely to recycle in order to obtain some commercial benefit from recycling or to save costs on purchasing a new item.

In Nigeria, Kaduna and Abuja are two of the states that are part of the ACCP. An assessment of waste management in ACCP member states (ACCP, 2019) reports that Abuja has a controlled dumpsite where compaction is carried out, while in Kaduna, an open dump system persists.

Nigeria has an estimated population of 206 million generating between 00.5 kg and 1 kg per person per day equivalent to 53,600–107,100 tonnes per day generated by the urban population of the country (ACCP, 2019). The waste management authorities receive support from UNIDO in the form of policy and institutional framework development and from JICA (Japan International Cooperation Agency) in the form of training. Nigeria has a solid waste management budget of between 150 and 500 million NGN. Although there is no system introduced for taxing solid waste, a subsidy exists for projects that provide interventions on waste management. Figure 10.12 shows some LAWMA workers collecting waste from households on a street in Lagos. Several plastic items can be observed in the collected waste. Sacks containing separate plastics and other re-sellable items are also observed. From discussions, it was gathered that this waste is sent off to a dumpsite in the Igando area. A closer dumpsite is located in the Oke Afa area (Fig. 10.13), however, this is closed and no longer receives waste.

Fig. 10.12 Collection of household waste from a street in Lagos, Nigeria. June 2022

Fig. 10.13 A view of the Oke Afa dumpsite (closed) in Lagos, Nigeria April 2022

Examples of Informal Polymer Waste Management in Lagos

This section looks at some polymer-based materials and how their waste is managed in some parts of Africa. It looks at abattoirs, local food processing waste, hair, from salons, orange peels, and other fruit peels, fish scales, fish waste (sold as pet food), and chitinous waste. Here, we can see some creative ways in which the informal sector utilizes waste from food processing, textile and garment production, beauty salons, and others. (Speak with market women—fish traders and fruit traders, etc..—ask how much waste they generate a day on average and how they dispose of the different wastes if they are aware of alternative use like composting medicinal like fruit peel, etc.)

Here, some observations were made of trading activities which made use of products which contained materials. Commodities like fish, meat, hair, fruits, and corn are traded either in raw form or processed form across Lagos. This section focuses on the informal traders in markets and roadside trading points. Over a period, the activities of the traders are observed, and the traders are also asked questions about their activities and how they dispose of their waste from their trading activities. The following section discusses five different case studies.

Reuse of Plastic PET Bottles in Lagos

In the city of Lagos in Nigeria, plastic PET bottle reuse is a common sight. Plastic water bottles are collected by waste pickers, washed, and then sold to informal traders of certain goods who reuse them for packaging their products. Roasted peanuts, palm oil, vegetable oil, herbal preparations, brewed drinks, engine oils, and liquid soaps are some of the products that are packaged in reused plastic bottles in Lagos. From observations, these are mostly between 70 cl and 1.5 liter water bottles, however, nonalcoholic beverage bottles are also reused. The reuse of these bottles by the informal market contributes to reducing the amount of plastic waste generated however concerns exist as to hygiene and health safety, especially when used for ready-to-eat food products like peanuts and beverages. For example, how well these bottles are washed before reuse and the level of contamination. Figure 10.14 shows examples of some roasted peanuts packaged with reused plastic PET bottles previously used for packaging drinking water. The figure also shows how old woven polypropylene sacks are used as covers for wooden display tables and polyethylene bags used as covers for boiling corn. This method is intended to keep in much of the steam for faster cooking.

Fig. 10.14 Roasted peanuts packaged with reused PET bottles in Lagos, Nigeria. May 2022

Fish and Snail Waste Management in Lagos Market

Markets across Lagos sell various species of fish in fresh or frozen states. Some are supplied by local fishermen who fish in the lagoon, while others are sourced from cold rooms owned by suppliers who source from fishing boats. From interactions with some fish traders in the Lagos area of the Ejigbo market, some traders sell the internal organs of fish to individual buyers who mainly use them as pet food. On average, a 5-liter bucket of catfish internal organs gets sold for around N500 as of March 2022. This is equivalent to just under 1 USD according to the estimated exchange rate for the period. Scaly fishes like tilapia are typically sold with the organs except where the customer requests to have the internal organs removed. The scales are almost always removed at the market. Usually, the descaling is done in such a way that the scales accumulate on the floor. These are swept off at the end of the market day and taken away by waste disposal services in the market. As discussed in an earlier chapter of the book on biopolymers, fish scales are a source of collagen a natural biopolymer. The fish organs are also a source of protein another natural biopolymer. Figure 10.15 shows an example of tilapia fish getting scaled and cut at the point of sale at a market in Lagos.

Similarly, snails that are either sourced wild or farmed are sold at markets across Lagos. Often the buyers require the snails to get removed from the shells. From discussions with the snail traders, the shells are simply discarded and collected by the waste trucks that serve the markets. On request, the shells and other shells from past sales were given upon purchase of snails. Shells of snails contain chitin alongside minerals (Oyekunle & Omoleye, 2019). Figure 10.16 is an image of snail shells being separated from the snail meat at the market.

It is important to understand how these polymer-containing wastes are generated either on a small scale or large scale. This information helps in developing efficient strategies for sourcing polymer waste from food processing and designing

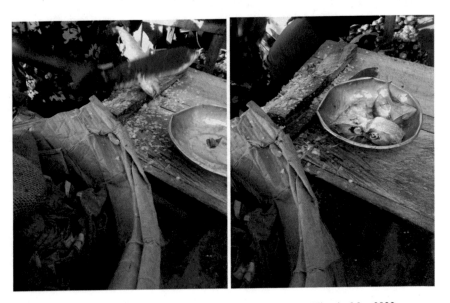

Fig. 10.15   Tilapia fish scales getting removed at a market in Lagos, Nigeria, May 2022

Fig. 10.16 Snail shells being separated at a market in Lagos, Nigeria May 2022

infrastructure such as a waste segregation system at markets where certain types of waste of value have dedicated bins and biopolymer manufacturers can have access to efficiently collect and utilize these currently underutilized resources.

Waste from Fruit Peels

Across the city of Lagos in Nigeria, vendors selling fruits, vegetables, and other fresh produce can be seen in different parts. Tropical fruits in particular oranges and pineapples are typically peeled at the point of sale. Whole pineapples are cut into sections that are sold individually in ready-to-eat plastic wraps which are mainly thin LDPE bags. These traders tend to accumulate a lot of these fruit peels daily. In making beverages such as a drink known as "zobo" which is brewed from an aqueous extract of the purple hibiscus flower, often fruit peels are added during the extraction to add flavor to the beverage. Individuals or commercial traders of the "zobo" drink obtain these peels from the fruit traders. Often in small quantities, these peels are given to buyers who have purchased fruits upon request. Larger quantities may be sold at the agreed price. Giving out or selling the peels relieves the fruit trader of the responsibility of handling the waste.

Pineapple peels are a source of lignocellulose which can serve in various applications. For example, in the production of lignocellulose nanocrystals, the pineapple peels are processed by hydrolysis with sulfuric acid. Lignocellulose nanocrystals get used as a stabilizer for Pickering emulsions (Chen et al., 2021). Figure 10.17 shows an image of a fruit trader in Lagos bagging some pineapple peels for a customer.

Opportunities in Waste Management in Africa

From the discussion so far, there are several opportunities in waste management in Africa. These include job creation, the building of enterprises around waste, improvement in tourism, improvement in health and well-being, freeing up land space for habitation and raising the value of communities, and improving crop yield and fisheries. Some of these opportunities are linked to abating the detrimental impact of improper waste dumping. For example, with the installation of well-engineered landfills, the deterioration of soil due to the seepage of dioxins and other toxins into the soil of nearby farms will be prevented. Similarly, the proper management of waste prevents contamination of aquatic environments which in turn improves fishing yield.

Currently, many Africans obtain their livelihoods from the waste business. For example, a study on the Kiteezi landfill in Kampala city of Uganda stated that on average, an informal waste worker earns an estimated 3.7 USD daily from waste picking at the Kiteezi landfill (Aryampa et al. 2020). Studies show that closure of the site will directly impact the livelihoods of the informal waste pickers therefore adequate support should be provided for them to mitigate against the hardship they may face upon closure of the landfill which serves as their main source of income. A well-planned waste management strategy can allow revenue to be generated from economic instruments such as taxes, dumpsite fees, and fines. These funds should then be directed towards the development of waste management infrastructures and reduction strategies.

The review suggests that there is a dearth of waste management services in countries like Kenya, Senegal, and Nigeria. This means that opportunities for new enterprises and job creation exist in waste management. Integrative waste management systems already exist in some countries which allow private companies to partner with the government to provide waste management services. In Dakar, for example, only one French company provides waste collection service in the city.

Fig. 10.17 A fruit trader bags some pineapple peels for a buyer. Lagos Nigeria. May 2022

The current data suggest that the existing waste management framework in most parts of Africa is not at optimum. The opportunity, therefore, exists for new technologies and operational models that can further improve waste management in this region. Multiple sectors are affected by solid waste management. These include tourism, healthcare, and manufacturing. The municipal solid waste generated in Africa contains between 70 to 80% recyclable contents (UNEP, 2018) such as plastics, metals, glass, and paper. Currently, only around 4% gets recycled, and mainly for the purpose of gaining profit. If well-managed Africa could recover an estimated 8 billion USD worth of value from waste annually. This will require a short-term investment of between 6 billion USD and 42 billion USD (UNEP, 2018).

Conclusion

It is important that the rate of development of waste management systems and technologies in this region must keep pace with the rapid rate of urbanization. Otherwise, with fast-growing cities and increased consumption and disposal of plastics and polymer-based products, the accumulation of waste generated from such consumption is likely to result in waste problems which lead to environmental and health problems.

With a rising population and more demand for land for habitation and agriculture, landfilling or open dumps become even less of a sustainable option. Yet, this is currently the prevalent mode of waste management across the African region. There is a need for the adoption of more sustainable waste disposal methods such as recycling and energy recovery. Milestones to achieving this include developing and implementing more viable waste management systems with well-implemented financing structures, training, and capacity building in waste management, and development, and maintenance of waste management infrastructures,

The risks associated with improper waste management include contamination of groundwater, degradation of the soil, and loss of spaces which could have otherwise been used for human habitation, agriculture, and commercial activities to waste dumping. Non-biodegradable plastics and polymers in particular pose a problem in dumpsites and landfills as they are likely to remain there for thousands of years if not processed.

As many parts of Africa continue to show increased urbanization, it is important to shift this development towards a more sustainable path. It is important to adopt more sustainable urbanization where the three Rs of sustainability are integrated into the lifestyle and waste management system. This will allow the African continent to benefit from sustainable urbanization rather than suffer the adverse impact of unsustainable urbanization.

There is insufficient data on waste management in Africa. Nonetheless, organizations such as UNEP, the world bank, African Union among others have provided some data which have been useful sources for this chapter. Improving information available on waste flow and waste management is important in developing facilities and capacities in waste management.

References

Abd'Razack, T. A. N., Medaiyese, S. O., Shaibu, S. I., & Adeleye, B. M. (2016). Habits and benefits of recycling solid waste among households in Kaduna, northwest Nigeria. *Sustainable Cities and Society., 28*, 297–306.

ACCP. (2019). *Africa solid waste management data book 2019*. Ministry of the Environment of Japan, UNEP, UN-Habitat

Aryampa, S., Maheshwari, B., Sabiiti, E. N., Bateganya, N. L., & Olobo, C. (2020). Understanding the impacts of waste disposal site closure on the livelihood of local communities in Africa: A case study of the Kiteezi landfill in Kampala Uganda. *World Development Perspectives., 25*, 100391.

Bappayo, A., & Maidunoma, Z. (2018). Role of radio stations in creating awareness on proper solid waste management practice in Yobe state, Nigeria. *Socioeconomic Challenges, 3*(2), 95102.

Beri, K. Y. V. (2018). Improvement of the waste management system in Senegal. *Mediterranean Journal of Basic and Applied Sciences., 2*(3), 105–126.

Chen, Y., Zhang, H., Feng, X., Ma, L., Zhang, Y., & Dai, H. (2021). Lignocellulose nanocrystals from pineapple peel: preparation, characterization, and application as efficient Pickering emulsion stabilizers. *Food Research International, 150*(PartA), 110738

German Federal Ministry for Economic Cooperation and Development. (2021). *Sector brief Senegal: Solid waste management and recycling.* Deutsche Gesellschaft fur Internationale Zusammenarbeit (giz) Gmbh.

Gwada, B., Ogendi, G., Makindi, S., & Trott, S. (2019). Composition of plastic waste discarded by households and its management approaches. *Global Journal of Environmental Science and Management.* https://doi.org/10.22034/GJESM.2019.01.07

Hoornweg, D., & Bhada-Tata, P. (2012). *What a waste: A global review of solid waste management.* World Bank. Urban Development Series Knowledge paper 15.

IPEP International POPs Elimination Project. (2006). *The waste management issue in Senegal: The example of Thies: outlining solutions to the waste incineration.* Association pour la Defense de l'Environnement et des Consommateurs (ADEC).

Kabera, T., & Nishimwe, H. (2019) Systems analysis of municipal solid waste management and recycling system in east Africa: benchmarking performance in Kigali city, Rwanda. In *E3S Web of Conferences* (vol. 80, p. 03004).

KCCA Kampala Capital City Authority. (2020). *Preliminary information memorandum Kampala waste PPP project.* IFC International Finance Corporation World Bank Group.

Oyekunle, D. T., & Omoleye, J. A. (2019). Effect of particle sizes on the kinetics of demineralization of snail shell for chitin synthesis using acetic acid. *Heliyon, 5*(11), e02828.

Selin, E. (2013). *Solid waste management and health effect: A quantitative study on the awareness of risks and environmentally significant behavior in Mutomo Kenya.* BSc. Thesis, UMEA Universitet, Sweden EnterpriseWorks/VITA.

Tomita, A., Cuadros, D. F., & Tanser, F. (2020). Exposure to waste sites and their impact on health: A panel and geospatial analysis of nationally representative data from South Africa, 2008–2015. *The Lancet, 4*(6), E223–E234.

UNEP. (2018). Africa waste management outlook. Summary for decision-makers.

Yoada, M. R., Chirawurah, D., & Adongo, P. B. (2014). Domestic waste disposal practice and perceptions of private sector waste management in urban Accra. *BMC Public Health, 14,* 697.

Chapter 11
Microplastics: Emerging Issues in Emerging Urbanization

Abstract Microplastics had been previously excluded from the data on global plastic pollution. Reports of findings from recent years have placed microplastics higher up the priority list. Microplastics have made their way into groundwater, soil, and the oceans. They have also been found to be present in the air samples. Thanks to studies from several research groups we can now identify the main sources of microplastics and their pathway to the world's ocean. While previous reports suggest that plastics released from synthetic textiles containing polyester were the leading source of microplastics, even more, recent studies that focused on microplastics released from paints and coating suggest that paints contribute more to the microplastics in the environment than even textiles. Parameters that affect the rate of microplastic release include washing methods, construction styles in case of release from painted buildings and structures, age of products like tires, climate, and weather conditions. These parameters vary across different regions and so does the impact of microplastic pollution and how it is addressed. Therefore, in this chapter, we will explore the issue of microplastics with a focus on the African region. It discusses microplastics in general as a global issue and a closer look at factors within the African continent that may influence the generation of microplastics from different sources in the region. In the process, the environmental and health implications of microplastics are explored as well as other issues.

Keywords Microplastics · Urbanization · Coagulation · Textiles · Polyester · Paints

Introduction

Plastics have come to be part of every aspect of modern life. They are so far one of the widely used classes of materials across the world. Plastics are used in diverse products that give durability, convenience, comfort, and practicality in various applications. In addition to the comparatively lower cost of production and processing, plastics have a high preference over other materials. Global plastic production as of 2019 was reported to be 368 million tonnes (Tiseo, 2021). A significant increase from the 350 million produced in 2015 (Geyer et al., 2017). Modern life has become almost totally dependent on plastics of one form or the other. Plastics are used in everything from the construction of buildings for homes, businesses, and institutions, footwear for fashion and sports to medical implants and personal care products.

The textiles industry, for example, was remarkably transformed by the innovation of incorporating plastics into fabrics. One of the most widely used plastics in fabrics today is polyester. Clothing items like T-shirts, socks, leggings, tights, undergarments, and many more, are a result of the ability to incorporate plastics like nylon and polyester to provide stretch, fit, style, and comfort that such garments offer. On the downside, these benefits of plastics in fabric do pose a heavy cost to the environment. Already the impacts are seen in the form of plastic pollution on land and in the oceans and waterways. Plastic pollution is already regarded as one of the greatest challenges of this era. Discovering that these plastics are now breaking down into microplastics that persist in the environment sheds new light on plastics.

Compared to larger plastic pieces and debris, microplastics are much smaller, have lower ease of detection, and higher likelihood of absorbing harmful microbes, metals, and toxins. This makes them a bigger threat to the environment and public health compared to macroplastics or larger plastic litters. It is estimated that every year 1.5 million tons of primary microplastics make their way into the world's ocean (Bouche rand Friot, 2017). Although around 98% of microplastics in the ocean originate from land activities, these make their way into the ocean through road runoffs, effluent from wastewater treatment, and wind. Only around 2% of the microplastics in the ocean originate from ocean activities like plastics in paints used in coating marine vessels. Until recently, the issue of plastic pollution has been dealt with from the angle of mismanaged waste. However, a large proportion of microplastics, in particular, primary microplastics originate from the process of using, manufacturing, or maintaining plastics and plastics products. For example, the plastic waste management approach will only consider a polyester garment as plastic pollution when the garment as a whole is disposed of. However, new understanding has revealed that microplastics are being shed from the garment from its production and throughout its use and maintenance. Every time the garment is washed and dried, it releases microplastics into the environment.

Recent reports have confirmed that microplastic release patterns significantly vary from region to region. This is expected since the use of plastics also generally varies by region. For example, plastic use in Africa is estimated at 16 kg per person per year while in Europe this is estimated at 140 kg per person per year in 2015 (Boucher & Friot, 2017).

What Are Microplastics?

Microplastics as the name implies refer to fragments of plastics having a size within the micron range? They are generally accepted to be plastic particles less than 5000 microns in size (Boucher & Friot, 2017). They either originate from larger plastic pieces or are intentionally prepared in this form for specific applications. The breakdown of plastics into microplastics is not synonymous with biodegradation. Microplastics still retain the chemical identity of the parent plastic piece. Polypropylene microplastics, for example, will remain a micron-sized particle of polyethylene. On the other hand, biologically degraded polyethylene has had its chemical structure altered by the process of degradation, and chemical bonds have been broken by the action of enzymes from microbes and/or UV radiation. The product of such degradation is, therefore, no longer polypropylene but has been broken down into other smaller compounds like methane, carbon, water, and hydrogen.

Whereas large plastic items like bottles, shopping bags, and straws are easily visible to the naked human eye thus making them easy to get picked up and easily distinguished and separated from other materials in the environment and waste stream, microplastics pose a serious challenge in this regard. Most commodity plastics that get used in everyday life density lower than or close to that of water. Most have densities between 0.85 and 1.41 g/cm3 (Eerkes-Medrano et al., 2015). So most of these plastics tend to float on water. For example, density of PET is 1.38 g/cm3, that of polypropylene is 0.92 g/cm3, while that of polystyrene is 0.909 g/cm3. These densities are in close range to the density of water which is around 1 g/cm3 and that of seawater only slightly higher varying between 1.02 and 1.03 g/cm3 with the colder regions of the ocean having a higher density than the warmer regions (Webb, 2021). This density factor in addition to buoyancy and surface forces of water means that microplastics will remain suspended in water. This is unlike nonplastic materials like clay which eventually settle due to their higher density. In addition to their low densities, these plastics are generally hydrophobic and therefore not likely to soak up or bond with water to form aggregates that become heavier and settle. Because plastics tend to repel water, this property further makes microparticles able to disperse well in water. This property is even desirable in products like facial scrubs which use microplastics.

Types of Microplastics

Microplastics are categorized according to various criteria. These include size, origin, shape, or plastic types. According to their origin, microplastics are classified as either primary or secondary microplastics. Primary microplastics are those that have been intentionally produced as microplastics by manufacturers. The plastics serve their function in this form. An example of this is the microplastics referred to as microbeads used in facial scrubs and toothpaste (Cheung et al., 2017). Secondary microplastics are those that occur as a result of discarded larger plastic objects breaking down by fragmentation and degradation in the environment into microplastics. These are in other words unintended microplastics. Exposure to ultraviolet radiation, abrasions, or physical and chemical stress in the environment facilitates the degradation of plastic debris into microplastics. Microplastics from textiles and car tires occur as a result of plastic fibers in textiles being fragmented and shed while still in use. Therefore, these mainly come in the form of microfibres (Roos and Levenston, 2017) and are also classified as primary microplastics (Boucher & Friot, 2017). Primary microplastics are microplastics directly introduced into the environment from a source, while secondary microplastics are those that are generated over time from discarded plastic pieces and debris already in the environment.

Microplastics are also categorized on the basis of the parent plastic type. Some plastic types are more common to a specific application. For example, microbeads in cosmetics were predominantly made of polyethylene before these were banned and polyester has been widely used in synthetic textile, although other plastics like polypropylene and polyamides are also used for textiles.

The most common commodity plastic types are assigned resin identification numbers 1–6 with the resin identification number 7 assigned to all other plastics. Plastic types 1–6 being; PET (polyethylene terephthalate), HDPE (high-density polyethylene), PVC (polyvinyl chloride), LDPE (low-density polyethylene), PP (polypropylene), and PS (polystyrene), respectively. The plastic type from which the microplastics originate significantly determines the behavior of the microplastic since the chemistry of each plastic varies. For example, how the microplastics get dispersed in the medium and the interaction with the other chemical or biological components of the medium such as air, water, or soil. Polyethylene, for example, comprises mainly carbon and hydrogen in its structure. This gives it a more hydrophobic property compared to nylon which has amide and carbonyl groups in its structure. These functional groups make nylon more hydrophilic than polyethylene. Therefore, when the microplastics of polyethylene and nylon are released into a water body, they tend to have different impacts to an extent. There is a myriad of plastic types from very diverse sources in the ocean and other water bodies this calls for effective means of sorting microplastics into different types in order to aid the study of the impacts and methods of processing each microplastic type as well as a combination of microplastic types.

The size and shape of microparticles can also be used as a basis to categorize them. Microparticles can be spherical, fibers (microfibers), and other regular and irregular shapes. The shape depends on the conditions under which the microparticles are formed and exist. Primary microparticles that are intensionally produced tend to be of regular shapes to suit their specific application. This is achieved through controlled and reproducible production processes. Out in nature where the parameters are variable, the shapes tend to be more irregular. In textiles where the parent plastic source is a fiber, the microplastics released are more likely to take on microfiber shapes.

The size of microplastics varies significantly from just a few micrometers in diameter to thousands of micrometers. Samples of microparticles are collected from a medium like a body of water (Bermudez and Swarenski, 2021) or air samples in a room (Tao et al., 2022). The microparticles can then be separated from macroparticles and in separate size ranges using sieves for particle size analysis. Microplastics are predominantly between 100 and 800 microns in size (Hermandez et al., 2017). Various sources define microplastic sizes differently, for example, some studies report microparticles as having less than a 2.5 mm radius (Gouin et al., 2015; Sillanpaa & Sainio, 2017). It is generally agreed that microplastics are those below 5000 microns

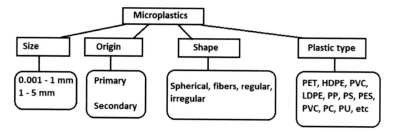

Fig. 11.1 Microplastic classification

(Boucher & Friot, 2017). A recent study proposes the size classification of micropar-ticles based on the size distribution of plankton. The motivation for this is that the size categorization method is in harmony with the biological system of size distribution that exists in nature and has been in use for over 120 years. Such that the microplastic range is between the size of bacteria and amphipods (Bermudez and Swarsenski, 2021). Since the term microplastics in general use includes all plastic fragments less than 5 mm in size, in the standard length scale this includes nanoplastics and macroplastics within this range. In the new plankton-based classification, therefore, the size included virus-size particles between 0.02 and 0.2 microns to amphipods and similar-sized planktons less than 5 mm in size. Figure 11.1 summarizes the different microplastics classifications.

Environmental and Health Impact of Microplastics

Even based on their size alone, microplastics are inherently challenging to deal with the additional challenges that are already posed by plastics make microplastics even more challenging than larger sized plastics. This is why the issue of microplastics has to be given separate consideration rather than being generally treated as a plastic waste management problem. There is existing evidence of microplastics, in the soil, water, and air we breathe (Boots et al., 2019; Romeo et al., 2015; Tao et al., 2022) and microplastics have made their way into the food chain and into human digestive systems (Yan et al., 2022). These discoveries are relatively new therefore there is much yet to be learned about the long-term impact of microplastics and the extent to which they have contaminated the environment and their distribution across different regions.

Microplastics can get into the human body through inhalation of air contaminated with microplastics or ingestion of food or water that is contaminated with microplas-tics. Microplastics have been found in the feces of humans and have been linked to inflammatory bowel diseases (Yan et al., 2022). When the stool of healthy people was compared to those of people with inflammatory bowel diseases, the results revealed that those who had inflammatory bowel diseases had significantly higher concen-trations of microplastics detected in their stool. Microparticles found in the stools were of 15 different types, however, the amount of PET and polyamide were higher. Microparticles have also been linked with pulmonary diseases based on experiments on cultured human lung cells which showed retarded cell proliferation and morphological changes triggered by the presence of polystyrene microplastics (Goodman et al., 2021). This suggests that the release of microplastics into the environment has a detrimental health impact. Already studies have reported deficiencies in diagnosis and inadequate clinical capacity in the management of inflammatory bowel diseases in sub-Saharan Africa (Watermeyer et al, 2020). The limited data available on the incidence of inflammatory bowel disease in sub-Saharan Africa suggests a rising incidence (Hodges & Kelly, 2020). An impending increased occurrence of diseases due to microplastics will mean more pressure on the already stretched healthcare system.

It is already well established that plastics in the ocean have gotten into the bellies of water animals, some of which have led to the death of these animals. The same has been discovered of microplastics, however, the impacts of microplastic ingestion by aquatic animals and humans are yet to be fully established and are more challenging to measure. This can be partly attributed to the fact that microplastics are hardly visible to the naked eye. Notwithstanding, based on existing knowledge, the impacts of microplastics on the environment are not expected to be less severe than that of larger plastics. Both large and small aquatic organisms depend on tiny microscopic organisms known as phytoplankton and zooplankton for food. Phytoplanktons, directly and indirectly, provide food for different organisms from small fishes to even giant whales and sea turtles. They, therefore, form a key part of the food chain. Since humans and other animals that are eaten by humans feed on these aquatic animals, these phytoplanktons are an important part of the food chain. Organisms that feed on these phytoplanktons tend to also take in the microplastics inevitably since they are of similar sizes (Romeo et al., 2015).

Even where the plastics are not reactive with water or many biological fluids such as intestinal fluid and blood, by virtue of being micron-sized indigestible, these microparticles can cause blockages and lead to impairment of the tissues and organs when ingested by animals (Romeo et al., 2015). The diffusion of light and other physical properties in the water may also get al.tered by the microplastics. This in turn has an impact on the aquatic ecosystem.

Out in the environment, microplastics can provide a surface for pathogenic microbes to attach and form biofilms (Shat et al., 2008). Metals can also be absorbed into microparticles from their surrounding environment. When ingested by living organisms these pathogens and metals and other potentially toxic compounds are ingested alongside the microplastics. Therefore, beyond their own chemical composition and physical presence, microplastics can facilitate the growth and spread of harmful microbes. Studies have sufficiently established the fact that biofilm and other organic matter get attached to microplastics and are transferred into the bodies of the aquatic organisms that ingest them (Farrell et al., 2013).

It has been shown that soil exposed to microplastics of PLA was less suited for the germination of seeds. Plants grown in soil contaminated with microplastics had reduced shoot height. The soil pH was also reduced by the presence of microplastics of HDPE. The presence of microplastics also affected the water-stable aggregates in the soil, a factor important for soil stability. Such findings suggest that even biodegradable microplastics from plastics like PLA poses some adverse impact on the environment (Boots et al., 2019). Microplastics have been detected in agricultural soil and groundwater in different parts of the world (Dong et al., 2022). These findings indicate that crop yield may be affected by the presence of microplastics in the soil. The contamination of groundwater with microplastics might have a more direct impact on communities where untreated well water is consumed for drinking and cooking.

Sources of Microplastics in the World's Oceans

Microplastics were previously not included in previous global plastics pollution data. The first global assessment to be published on sources of microplastics was in 2017 (Boucher & Friot, 2017). This exclusion of microplastics from plastic pollution estimates was later found to significantly underestimate the amount of plastic in the environment by millions. These microplastics are a result of plastic pieces breaking down over time into micron-sized fragments. These come from synthetic textiles made from plastics like polyester, car tires, plastics suspended as dust particles, plastics in road markings that wear off over time, microplastics in personal care products used as exfoliants, plastic pellets from plastic product manufacturing, and plastics used in coatings. Note that although the general term "microplastics" is adopted, these also refer to other polymers like the rubber in car tires.

Although only around 2% of the plastics in the sea originate from sea activities and 98% originate from land (Boucher & Friot, 2017), the microplastics released on land remain for at least 15 years before being washed into the sea (Zubris et al.,2015). Across the world, around 69.7 million tons of fiber are consumed for apparel production every year. Around 60.1% of these fibers are synthetic fibers (FAO, 2013). More of the fibers produced globally are increasingly synthetic fibers and around 62.7% of these are consumed in developing countries including countries in the African continent. The amount of textiles consumed per person on average in Africa is less than that consumed in, for example, North America. While it is reported that the average American consumes around 17.8 kg of textiles per year, the average African is estimated to consume 0.7 kg (FAO, 2010).

Tires are said to be the second highest generating source of microplastics according to Table 11.1 (Boucher & Friot, 2017). There are an estimated 1.413 billion vehicles in use in the world as of 2010 (Boucher & Friot, 2017). Eighty-four percent of the vehicles in the world are in Asia, Europe, or North America (36%, 27%, and 21%, respectively). This means a small fraction of vehicles in Africa compared to the developed economies. Around 57% of the synthetic rubber produced in the world went into tire production in 2010; 13.9 million tons of tires get sold every year on average, 46% of this was synthetic rubber.

Table 11.1 Sources of microplastics in the ocean according to the International Union for Conservation of Nature

Microplastic source	Percentage in the world's oceans (%)
Synthetic textiles	35
Car tires	28
City dust	24
Road markings	7
Marine coatings	3.7
Personal care products	2
Plastic pellets	0.3

More recent studies suggest that microplastics from paints are a priority concern to the environment (Paruta et al., 2021). Paints are used in architecture, marine, road construction, automotive, general industrial applications, furniture, and many other applications. Over the years, these paints gradually wear off. Around 37% of paint is polymer materials. The IUCN estimates the global leakage of paint to be between 5.2 and 9.8 million tonnes annually. An estimated 37% of this (1.9 million tonnes) ends up as microplastics in the world's waters and 63% remain on land. Although this was not included in the 2017 report, recent data suggests that paints contribute even more to the microplastics in the ocean than synthetic textiles.

A previously ignored and underestimated issue, microplastics have moved up the priority list in recent years. Address this requires efforts in understanding the sources of microplastics and implementing reduction strategies. Such reduction strategies include the use of biodegradable fabrics, technologies to recover microplastics from the environment, and biodegradable coatings and paints. Table 11.1 lists the various sources of microplastics in the ocean according to International Union for Conservation of Nature in 2017 (Boucher & Friot, 2017).

Microplastics by Region

Africa and the Middle East contribute 8.7% to the global release of microplastics into the ocean (Boucher & Friot, 2017). These countries account for the lowest when compared to other regions of the world. India and South Asia contribute 18.3%, North America contributes 17.2%, Europe and Central Asia combined contribute 15.9%, China contributes 15.8% East Asia, and Oceania contributes 15%, while South America contributes 9.1%. The regions are faired here based on the oceans surrounding them such that the contribution of microplastics into the ocean for some paired regions might not be easily distinguishable from one another.

These figures are based on estimates which measure microplastic impact based on population, affluence, and technology efficiency in each region. Factors influencing microparticle release into the ocean from each region in terms of technical efficiency include wastewater treatment systems available. For example, in a highly populated region with limited access to wastewater treatment, the microplastics released from this region are likely to be higher than in one with a lower population and better wastewater treatment facilities. However, where there is higher affluence, the population is likely to have a stronger purchasing power and more microplastics generated from purchased goods like textiles and vehicle tires. Therefore, in these estimates, we see, for example, even where Africa has a higher population and lower access to wastewater treatment systems than North America, a larger share of the population of North America has higher purchasing power and therefore is likely to purchase more products releasing more microplastics. Hence, based on the combination of these factors countries in North America release more microplastics into the ocean compared to those in Africa based on this analysis.

In Africa and the Middle East, more of the microplastics released are from synthetic textiles washing and tires abrasion on roads. In North America, Europe, and Central Asia, for example, abrasion of car tires on road contributes higher than textiles washing to microplastics release. In most of the more economically developed countries, washing machines with filters and wastewater treatment reduce the change of microplastics getting into the ocean. Lower income countries within the Middle East and Africa contribute around 12% of total world plastic leakage from paint into the world's ocean. This makes it the second largest contributor globally. Asia Pacific is the top contributor (Paruta et al., 2021). Figure 11.2 shows the microplastic contributions from different regions to the world's ocean.

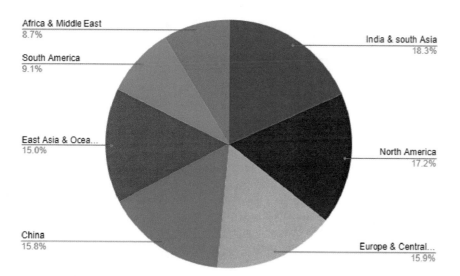

Fig. 11.2 Pie chart summarizing the contribution of different regions to microplastics leaking into the world's oceans

Microfibers from Textiles

A 2017 global assessment of microplastics in the world's oceans and waterways concluded that textiles were the leading source of microplastics in the ocean (Boucher & Friot, 2017). This is mainly from microfibers shedding during the washing of clothes and other textiles. Plastics in textiles over the years have given the advantage of increased comfort, ease of processibility, mass production, and lower cost compared to natural fibers like cotton and silk. Some synthetic plastic textiles have come to replace some natural fibers in products that were initially made from natural fiber textiles. For example, nylon replaced silk for hosiery. However, synthetic textiles which make use of plastic fibers like polyester are also associated with detrimental effects resulting from the release of plastics and microplastics into the environment. Although cotton textiles also release microfibers at a higher rate than polyester (Zambrano et al., 2019), these are biodegradable and will not pose a level of concern to the environment and health as synthetic non-biodegradable fibers. Microplastics released from sources depend on various parameters like the number of washing cycles, the temperature of washing, use of detergents, types of yarn, and finishing of the fabric. Understanding how these parameters relate to the release of microplastics from textiles can guide the government and other stakeholders in proposing policies and actions directed at managing microplastics released from textiles washing.

Microplastics are defined as plastic fibers and particles of less than 5 mm in size (Fontana et al., 2020). Microfibers from acrylic fibers, for example, have been reported to vary between $2411 \pm 1500 \, \mu m$ long and $18 \pm 4 \, \mu m$ in diameter (Mahbub and Shams, 2022). An estimated 60% of fibers produced globally are synthetic fibers. Of this, polyester and polyamide make up the largest portion (Fontana et al., 2020). In 2020, global polyester production was reported to be 57.1 million metric tons (Statista, 2021). Applications include clothing, carpets, upholstery, and other applications.

In a single wash, a pair of jeans, for example, can release between 51,900 and 60,100 microfibers (Tao et al., 2022). With effective sewage treatment, much of these gets collected in the process of treatment and can be prevented from getting into the oceans and waterways. Poor operations, uncontrolled inputs, power cuts, and insufficient funding are part of the issues that have been reported on wastewater treatment in some African countries (Josiane et al., 2013). Therefore, the public wastewater treatment can't be relied on to prevent the microplastics from municipal activities like textiles washing, from getting into the ocean and other parts of the environment in this region.

When cotton, rayon (regenerated cotton), and polyester knitted fabrics were compared in an accelerated laboratory simulated washing and home washing, all released microfibers, however, cotton fabrics released more microfibres. The amount of microfiber released was increased by increasing the temperature of the washing and the use of detergent. Cellulose fabrics release between 0.2 and 4 mg per gram of fabric, while polyester fabrics release between 0.1 and 1 mg per gram of fabrics in accelerated laundering experiments (Zambrano et al., 2019). In other studies, under varying washing conditions, up to 38.6 mg per kg of fabric has been reported for 100% polyester fabrics (Fontana et al., 2020).

The properties of the fabric in terms of the yarn strength, abrasion resistance, and hairiness also determine the rate of microfibre release from the fabric. To reduce the rate of microplastics from fabrics, they should be made with stronger yarns, less hairiness, and higher abrasion resistance. Therefore, generally, higher quality fabrics are less prone to the shedding of microplastics and microfiber. Despite cotton fabrics being biodegradable, the release of these microfibers into the environment can have other effects such as increasing the oxygen demand and microbe balance.

Other than the microfibers and microparticles released from the washing process, the drying process also releases microplastics into the environment. This is exacerbated when tumble dryers are used. In a 15-min wash, a tumble dryer can release between 433,128 and 561,810 microfibers into the surrounding air (Tao et al., 2022). This can be mitigated by the use of a filtration system within the dryers and washing machines. On average, Canadian homes release between 90,000,000 and 120,000,000 microfibers from a dryer every year (Tao et al., 2022).

The older a garment is, the fewer microplastics get shed during washing. Studies on acrylic fabrics, for example, showed that microfibers released from the first wash were 45% less than the amount released from the seventh wash (Mahbub and Shams, 2022). Much of the microplastics is released from the first few washes. The world's second-hand clothing market is valued to be over 1 billion USD as of 2005 (Baden & Barber, 2005). Second-hand clothing provides clothing at reduced cost and keeps the textiles away from the waste stream for longer. On a global scale, second-hand clothing accounts for less than 0.5% of the total clothing value. However in sub-Saharan African countries imported second-hand clothing once made up around 30% of the total value of imported goods and over half the volume of imported goods according to Oxfam reports (Baden & Barber, 2005). With increasing trade with Asia, the second-hand clothing market in sub-Saharan Africa is being overtaken by newly made cheaper clothes. More of these are likely to be made from synthetic fibers since these have lower costs compared to natural fibers like cotton. Countries in the EAC are taking steps to stop the importation of second-hand clothing in the countries. This is in an effort to grow the domestic textiles industry. Since the textile industry is more labor-intensive, particularly with more natural fibers like cotton which involve agricultural input, it requires less complex facilities and therefore a promising path to industrialize these countries. In Rwanda, for example, in 2016, applied a higher duty on the importation of second-hand clothing and shoes (USAID, 2017).

The value-added services to used clothing include transportation, cleaning, repairing, and sometimes restyling (Hansen, 2014). The energy inputs and emissions from these additional processes need to be considered to determine the sustainability of exporting used clothing across the oceans. On the other hand, when the clothes are discarded in the environment they are likely to break down into microplastics.

Washing for a longer time, use hot water, and use detergents. For example, twice as many microfibers get released from acrylic fabrics when the washing time is increased from 30 to 60 min. The microfiber released from acrylic fabrics increases from 60.22 + 13.32 mg/kg without detergent to 162.49 + 44.21 mg/kg when detergent is used. When the temperature is increased from 20 to 40 oC, the microplastics released increase by 1.8 times. In many parts of Nigeria, for example, handwashing is generally done with water at room temperature and drying is more commonly done in the open air on drying lines, rails, grass, or other surfaces rather than using tumble dryers. However, with the growing middle class and increased urbanization, more Africans are likely to adopt the use of washing machines.

There are significantly fewer paved roads with road markings in Africa compared to other regions like Europe and North America (Gwilliam et al., 2008). Considering the income of this region, the cost of road maintenance and construction is a significant chunk of its GDP. The asset values in some sub-Saharan African countries like Malawi, Niger, and Madagascar are over 30% of their GDP (Gwilliams et al., 2008). Major road networks that are used to transport goods of around 200 billion USD annually could be no more than 10,000 km. Maintaining the roads to good enough standards for sustainable transportation is challenging therefore the addition of paints for aesthetics and markings is often not used as much as in more developed countries. To connect the countries within sub-Saharan Africa to a significant level, 60,000–100,000 km of roads are required. World bank report states that less than 40% of Africans living in rural areas have roads that are passable all season within 2 km distance (Gwilliams et al., 2008). With recent years seeing accelerated rates of urbanization in the continent, significantly more road networks are being constructed across different countries in Africa. It is therefore important to seek ways to adopt sustainable practices that will ensure this increased road construction does not result in an increased rate of microplastics being released into the environment from the paints used.

Plastics in Synthetic Textiles

Natural fibers like cotton and hemp lack some properties desired in modern fabrics. Therefore, some fabrics in part or full make use of plastics fibers such as polyester and nylon to improve properties such as stretchiness, moisture repellent, form-fitting, texture, and appearance. In some cases, plastics make the fabrics more affordable. Plastics have been long used in the modern textiles industry (Giles, 2009).

Plastics are incorporated into textiles in a variety of ways such as coating plastics on fabrics, impregnated into fabrics made from other natural fibers like cotton, or spinning the plastics into fibers. Plastics are made into fibers using techniques like wet spinning and melt spinning. These are then converted into textiles used for making garments, bags, medical protective gear, swimwear, sporting gear, and several other applications. Indeed the ability to make plastics into textiles has significantly contributed to improving the quality of safety in health and protective apparel.

Polyester, polyurethane, and polyamides are examples of microplastics from textiles. The amount of microplastics released from washing textiles surpasses those from products like facial scrubs and toothpaste. Microplastics from textile washing also have a different pathway to the ocean from these other sources. Polyester is the second most widely used fabric today with cotton being the most widely used. Around the world, it is estimated that 46 million tonnes of polyester are produced annually (Carmichael, 2015). Every wash around 210,000 microparticles get released from a piece of polyester fabric (Sillanpaa & Sainio, 2017). The amount of microplastics released per wash varies. For example, in a different study, just over 1900 microplastics were released per wash. Various research studies have presented different values for the microplastics released from textiles washing. The results vary with different methods. Between 0.033 and 0.3% weight of microplastics get shed from polyester fabrics during the first wash (Hartline et al., 2016).

As the population rises, inevitably so does the number of textiles that get used and washed and so does the number of microplastics that get released into the environment, if the current usage pattern persists. The lower cost of synthetic fabrics also means more of this is produced to meet demand, especially in lower income countries. Finland, for example, releases around $1.54 \times 10^{\wedge}19$ kg of microplastics from textiles washing using washing machines in homes (Sillanpaa & Sainio, 2017).

Factors that Affect Microplastics Release from Textiles

In order to take actions to prevent the release of microplastics into the environment, the sources and the causes of the microplastic release from each source need to be well understood. A quantitative relationship between measurable parameters and the rate of release of microplastics from textiles has been part of the research efforts on microplastics, based on a review of studies that have been carried out on microplastics released from textiles in recent years. Figure 11.3 summarizes the factors that affect microplastic release from textiles based on findings from different studies (Hermandez et al., 2017; Falco et al., 2018; Lant et al., 2020).

Thus far, the use of detergents has been shown to have the strongest influence on the rate of microplastics released during the washing of textiles. When a liquid or powdered detergent gets used during the wash and all other factors are kept constant, the amount of microplastics released from the fabric is more than when no detergent gets used. For example, when detergents were used in washing a polyester fabric in one study, 0.1 mg of microfiber was released per gram of fabric. When no detergent was used under the same washing conditions and textile type, 0.025 mg of microfiber was released per gram of fabric (Hermandez et al., 2017). Already, detergents have some adverse impacts on the environment when released into the oceans and waterways.

How the fabric is constructed also determines the rate of microplastic release. Woven polyester fabrics will release more microplastics than knitted polyester fabrics under the same washing conditions (Falco et al., 2018). Using softeners and bleaches can also reduce the number of microplastics released in a given wash by up to 35% according to a study where 6 million microplastics got released from 5 kg of polyester

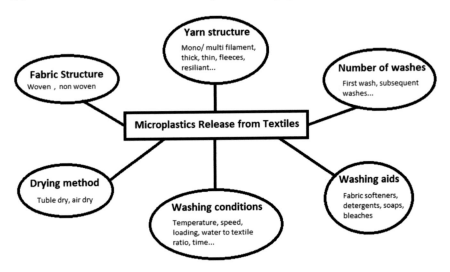

Fig. 11.3 Illustration of factors affecting microplastics release from textiles

fabric during a wash (Falco et al., 2018). Increasing the wash cycle time, temperature, loading, and intensity all lead to increased shedding of microplastics from the fabric. This can be attributed to the fact that these parameters are related to increased abrasion and weakening of the fabric.

The tendency of a fabric to shed microplastics during its usage and maintenance can be reduced by taking certain precautions during the production process. Woven fabrics release fewer microplastics than knitted fabrics. Making the yarns more resilient and larger also reduces the chances of microplastic release. Laser cutting of the fabric during production rather than cutting with scissors can also reduce microplastic release from it (Roos & Levenstam, 2017). Fabrics with loose knittings and fleeces are more prone to shedding microplastics, while the use of fabric treatments such as electrofluidodynamic method or pectin-based finishing formulations can reduce microfiber release (Lant et al., 2020).

Pathway from Textiles into the Environment

Across the world washing textiles is an essential activity that takes place daily on a large scale. Textiles also provide one of the basic needs of life; clothing. To some extent, it also gets used to provide shelter, another basic need. For example, textiles are used in make-shift tents and sheds. Microplastic released from textile washing is a major source of microplastic release into the environment. In more industrialized parts of the world, washing machines are routinely used for washing, and the wash water from these washing machines is directed towards a wastewater treatment facility before being released into the environment. Some of the washing machines can also be fitted with filters. In such cases, a significant amount of microplastics can be removed before the water is released into the environment. However, where washing is done outdoors and the water is released directly into waterways like lagoons and rivers, this directly releases large amounts of microplastics into the environment.

In the basic sewage treatment, methods used include biological digestion, drying, or lime stabilization. Once treated the waste is considered safe enough to be released into the environment. This can be on land or into the sea with the aim that the residue will be broken down by biodegradation. However, since microplastics don't degrade for over half a millennium, they persist and eventually, if on land, get washed into the aquatic environment through run-offs during rainfall (Mahon et al., 2016).

Analysis of solid residues from wastewater treatment plants in Ireland revealed that the solids from waste treatment plants that were regarded as safe to be released on land still contained 1,196–15,385 microplastics per kg of dry weight (Mahon et al., 2016). It was also found in the analysis that the size of microplastics released varied with treatment methods. Lime treatment resulted in size microplastics. Although this relationship between the treatment method and microplastic size in the solid residue is not fully understood, it is hypothesized to be the shearing of the microplastics by the lime. The study also suggested that lower microplastic content was also detected in anaerobically digested sewage sludge. Therefore, there is potential for improving wastewater treatment methods to minimize the release of microplastics.

Microplastics in the Aquatic Environment

Over 150 million tons of plastic are estimated to be discarded in the world's ocean (EPA, 2016), and between 0,1 and 1.5% of these plastics are estimated to be microplastics (Gouin et al., 2015). Microplastics get into the aquatic environment through various paths. Because microplastics can be made up of any of the existing plastic types, this means in any given sample of microplastics, the chemical composition can significantly vary. Since the densities of plastics vary, the composition of microplastics based on plastic types is expected to vary at different depths of the ocean.

The depth of the ocean certain types of microplastics settles depends on variables such as the shape of the microplastics, the salinity of the water body which in turn affects the density slightly, biofilm formation, and other factors. Available data suggests that microplastics are predominantly deposited in the parts of the oceans and waterways that are in close proximity to urban and densely populated land areas (Corsi & Mason, 2016). Some of these tend to be from plastics that have either been intentionally discarded or get washed up by rain and wind from land into the aquatic environment. These then eventually break down through weathering and abrasions into microplastics. Microplastics in the aquatic environment can therefore be a mix of primary microplastics from paints, textiles washing, and tire abrasion or secondary plastics from degraded macroplastics. Sources of microplastics in the aquatic environment can, therefore, vary from PET bottles, plastic plastics in footwear, agricultural mulch plastics, and as recent data now suggests predominantly, plastics from paints and washing of textiles (Laitala & Klepp, 2019). This variation in the sources and composition of microplastics poses certain challenges in the collection and processing of microplastics from the environment.

Microplastics disposed of in the aquatic environment are of particular interest because the oceans and waterways connect the world. Once microplastics are discarded in the ocean there is an increased tendency to spread. The aquatic environment being a continuous medium, the impact of microplastics such as acting as a surface for microbial growth is easier to spread unlike on land where such as is more restricted. Various figures have been presented on the extent of microplastics present in the environment. A study on microplastics in Germany reports that on average 0.0007 microplastic particles per liter of water in the groundwater (Mintenig et al., 2019). Another study on three Gorges reservoirs reports 4.7 particles per liter of water (Di and Wang, 2018). On average, 0.00026 microplastics per liter of water have been reported in Western Lake Superior (Sillanpaa & Sainio, 2017). Microplastics have been detected in treated wastewater (Iyare et al., 2020). Although the removal rates can be up to 94%, considering the large volume of wastewater that needs to be treated daily, this means microplastics are being released into the environment constantly. Increased awareness of microplastic impact has motivated research into improving the efficiency of microplastic removal and testing in wastewater treatment facilities.

Microplastics from Paints, Road Markings, and Coatings

Paints, marking, and coatings are used in large quantities in buildings, automobiles, road markings, anti-fouling, or anticorrosion paints on marine vessels, and other applications. Plastics in solution, thermoplastics, preformed polymer tapings, and epoxies are used as road markings. Coatings often are polymers dissolved in a solvent which is then spread or sprayed unto a surface to dry. Plastics such as polyurethane and epoxies are used for this purpose.

These paints, marking, and coating release microparticles when they experience abrasion. This can be from wind over time, rain, during cleaning, maintenance, or surface pretreatment for re-coating. Microparticles from paints contain more additives than other microplastics. This makes them likely to have a higher adverse impact on health if ingested or inhaled.

With increasing urbanization, more buildings are being constructed as well as more roads and other infrastructures such as bus and train stations across cities in Africa. Already the world consumes an estimated 1200 kilo tonnes of coatings used for traffic road markings (Grandview, 2016). The global demand for paint as of 2019 was estimated to be 52 million tonnes and 19.5 million tonnes of this included plastics (Paruta et al., 2021). Paints typically comprise about 50% polymers which are used as binders to hold pigment, extenders, or additives together (Lassen et al., 2015). Acrylic paints, for example, comprise 32% polymer binders, 6.5% pigments and additives, and 41% water. When the paint dries off, this leaves a polymer composition of around 60 to 70% volume (Iscen et al., 2021). Other binders include alkyd, polyurethane, epoxy, and chlorinated rubber (Turner, 2021). Currently, Europe and the United States are leading in this market (Boucher & Friot, 2017) but with increasing urbanization and more modern homes being painted, the market in Africa is expected to increase in the coming years.

The real estate industry in Africa is also projected to continue to grow significantly in the coming years. With expected net annual returns of around 20% in real estate investment in commercial buildings such as malls and office complexes, the sector continues to draw investors into the continent. The expected returns from real estate in Africa are higher than that in other continents (PWC, 2015). Although this is expected to see more uses for paints, buildings in new cities such as Eko Atlantic City in Nigeria, Konza City in Kenya, and Roma Park in Zambia have embraced the idea of sustainable buildings. The new awareness of microplastics from paint should drive the real estate sector towards further embracing the use of sustainable materials and alternatives to paints containing plastics and/or measures to mitigate microplastic leakage from buildings.

Between 5.2 and 9.8 million tonnes of plastic from paint leaks into the environment every year. From this, around 1.9 million tonnes of microplastics leak into the ocean and waterways of the earth. Paint leakage from architecture is thought to be the biggest contributor to global microplastic leakage from paint (Paruta et al., 2021). These new findings make paints a leading source of microplastic leakage into the ocean and waterways. As of 2017, textiles were thought to be the leading source of microplastics (Boucher & Friot, 2017). This has been previously overlooked.

Observations across Lagos, Nigeria, and from visits to other cities in Africa like Kigali and Accra, show most of the buildings in the cities are painted with common sightings of painted buildings and road markings fading over time. There is a mixture of constructed roads and earth roads, with much of the roads being earth roads. This is in accordance with the data showing fewer microplastics contributed from the African region (Fig. 11.2). Buildings and road markings are evidently less frequently repainted as frequently as they are in regions like Europe. Thus, indicating less chance of microplastic emission from paints. Some buildings with paints faded off as well as fading road markings in Lagos and Dakar are shown in Figs. 11.4 and 11.5. Figure 11.4 also shows some clothes, some of which are polyester fabrics, left air drying on lines outside. This method of laundry drying is a common sight. Although as the region rapidly becomes more urbanized, it is expected that more homes make use of tumble dryers.

Microplastics Removal Methods

The existence of microplastics in the environment and their sources is an emerging area of research and development. Recent efforts have been directed at understanding

Fig. 11.4 Images showing fading paints on a building in Lagos. The image also shows some polyester textiles dried out on laundry lines to air dry in the sun. Jakande Estate, Lagos, Nigeria, June 2022

the scale of microparticle release into the environment and the mechanisms by which they are released. Examples of these are experiments on the influence of washing conditions on the release of microplastics from textiles during washing cycles (Falco et al., 2018; Sillanpaa & Sainio, 2017). There are also research studies to quantify how much microplastics have been released into the environment using various measurement and analysis methods (Gouin et al., 2015; EPA, 2016; Boucher & Friot, 2017). The findings are then used to develop and implement approaches to manage the microplastics. Brand names in the textiles and garment industries like H&M are also funding research towards understanding the challenge of microplastics in the environment (Roos & Levenstam, 2017).

At present microplastics are not included in the regulations on wastewater treatment from effluent in textile industries (EC, 2009; Mahon et al., 2016). Textiles companies are required to carry out testing on their effluent, however, there is no regulatory requirement to test for microplastics at the time of writing this chapter. Coincidentally and fortunately so, some methods for the treatment of wastewater also remove microplastics. Standard wastewater treatments like anaerobic digestion or sublimation with lime will not eliminate microplastics from wastewater. Therefore, microplastics will only be removed where more advanced wastewater treatment methods are used. Regulatory requirements on wastewater treatment vary

Fig. 11.5 Fading road markings pictured in Dakar, Senegal, December 2019

from country to country therefore only in regions where strict regulations are implemented will microplastic be likely to be reduced significantly during the treatment of wastewater. Where there are ineffective facilities or weak enforcement of regulations, significantly more microplastics will get into the environment. The interconnected nature of the world's waters means microplastics being discarded in one region will eventually reach the ocean and therefore affects the global aquatic ecosystem.

Wastewater treatment methods like ultrafiltration and flotation can remove microplastics from wastewater effectively. However, in the low- and middle-income countries where wastewater treatment is basic or inefficient, less effective methods are employed. In such areas a mixture of treated and untreated wastewater enters the ocean and even when treated, it is unlikely that microplastics get removed (UNICEF, 2019; WHO, 2019). Analysis estimate that a single treatment plant releases around 65 million microplastic into the aquatic environment (Murphy et al., 2016).

Various innovations have been introduced to address the release of microplastics into the environment. For example, new washing machine designs include "eco wash" functionalities that wash at the optimal conditions that minimize the release of microplastics in the environment. There are new products in the market which claim to collect microplastics released from washing. Washing machines are also being designed with filters to remove microplastics from the effluent wash water before being released into the wastewater stream.

Alternatives to plastic microbeads used in cosmetics and personal care products have been developed. Examples of such are chitosan-based microbeads that are intended to replace polyethylene-based microbeads used in applications such as toothpaste, body scrubs, and shampoos (Ju et al., 2021). These products have demonstrated the processibility of chitosan into uniform-sized microbeads with the required mechanical properties and stability to serve the intended applications as effectively as microplastics but with the advantage of biodegradability.

Microplastics in water form very good suspensions that are supported by steric stabilization. One of the approaches to removing solid microparticles from a medium within which they are suspended is to coagulate the solids by destabilizing the suspension. This is done by adding electrolytes. Unlike electrostatic stabilization, steric stabilization is not easily lost by adding electrolytes to the suspension. Therefore, in order to separate microplastics from a water body, more advanced methods are required. Electrocoagulation has recently been applied for the removal of microparticles from wastewater (Shen et al., 2022). This method was shown to remove up to 98.4% of microplastics at optimal conditions with aluminum anodes proving more effective compared to a ferrous anode. The microplastics tested included PP, PMMA, and PE. It was also discovered that microparticles in the form of microfibers were more effectively removed than the microplastics in granular forms.

Another study showed that polyethylene microplastics can be removed from water with more than 99.2% efficiency using electrocoagulation (Perren et al., 2018). The process involves the neutralization of surface charges of the plastics using the process of electrolysis using a sacrificial anode and a current density of around 11amps/m^2. Electrocoagulation is a well-established method; therefore, it can be easily adapted for microplastic removal. The aggregation of the microparticles is caused by the metal ions that get released. These result in the formation of froths and sediments which are more easily separated from the water than microplastics.

Conclusion

The issue of microplastics brings a different dimension to a sustainable plastic economy. Where microplastics are not considered, plastic pollution is viewed as a waste management problem. Including consideration of microplastics reveals that a sustainable plastic economy should include the design of the product to the smallest detail. Understanding that every plastic and polymer product produced will ultimately end up in the food chain will motivate innovations towards truly environmentally friendly products. It also shows how efficiency in the system such as wastewater management is important in safeguarding health and protecting water and land resources. Assessing the dynamics of microplastics in different regions helps to identify the stakeholders and understand how various factors play out. This is important in creating tailored fact-driven solutions that are best suited to each region.

References

Baden, S., & Barber, C. (2005). The impact of the second-hand clothing trade on developing countries. Oxfam, pp. 1–35.

Bermudez, J. R., & Swarzenski, P. W. (2021). A microplastic size classification scheme aligned with universal plankton survey method. *MethodsX, 8*, 101516.

Boots, B., Russell, W. C., & Green, D. S. (2019). Effects of microplastics in soil ecosystems: Above and below ground. *Environmental Science & Technology, 53*(19), 11496–11506.

Boucher, J., & Friot, D. (2017). Primary microplastics in the oceans. IUCN International Union for Conservation of Nature. Gland, Switzerland. ISBN 978-2-8317-1827-9.

Carmichael, A. (2015). Man-made fibers continue to grow. *Textile World, 165*, 2588–2597.

Cheung, P. K., & Fok, L. (2017). Characteristics of plastic microbeads in facial scrubs and their estimated emissions in mainland China. *Water Research, 122*, 53–61.

Corsi, S. R., & Mason, S. A. (2016). Plastic debris in 29 Great Lakes tributaries: Relations to watershed attributes and hydrology. *Environmental Science & Technology, 50*, 10377–10385.

Dong, S., Yu, Z., Huang, J., & Gao, B. (2022). Fate and transport of microplastics in soils and groundwater. IN: Emerging Contaminants in Soil and Groundwater Systems: Occurrence, iMpact, Fate and Transport. Gao B (Ed). Elsevier. ISBN 9780128240885, pp 301–329.

EC. (2009). Directive 2009/28/EC of the European Parliament and of the Council of 23 April 2009 on the promotion of the use of energy from renewable sources and amending subsequently repealing Directives 2001/77/EC and 2003/30/EC.

Eerkes-Medrano, D., Thompson, R. C., & Aldridge, D. C. (2015). Microplastics in freshwater systems: A review of the emerging threats, identification of knowledge gaps and prioritization of research needs. *Water Research, 75*, 63–82.

EPA. (2016). Plastic microbeads in products and the environment. EPA: Sydney.

Falco, F., Gullo, M. P., Gentille, G., Pace, E., Cocca, M., Gelabert, L., Agnesa, M., Rovira, A., Escudero, R., Villalba, R., Mossotti, R., Montarsolo, A., Gavignano, S., Tonin, C., & Avella, M. (2018). Evaluation of microplastic release caused by textile washing processes of synthetic fabrics. *Environmental Pollution, 236*, 916–925.

Farrell, P., & Nelson, K. (2013). Trophic level transfer of microplastic: Mytilus edulis (L.) to Carcinus maenas (L.). *Environmental Pollution, 177*, 1–3.

Fontana, G. D., Mossotti, R., & Montarsolo, A. (2020). Assessment of microplastics released from polyester fabrics: The impact of different washing conditions. *Environmental Pollution, 264*, 113960.

Gwilliam, K., Foster, V., Archondo-Callao, Briceno-Garmendia, C., Nogales, A., & Sethi, K. (2008). Roads in Sub-Saharan Africa. World Bank and the SSATP. Paper 14.

Geyer, R., Jambeck, J. R., & Law, K. L. (2017). Production, use and fate of all plastics ever made. *Science Advances, 3*(7), e1700782.

Giles, E. V. (2009). Plastics and the textile industry. *Journal of the Textile Institute Proceedings, 40*(80).

Goodman, K. E., Hare, J. T., Khamis, Z. I., Hua, T., & Sang, Q. X. A. (2021). Exposure of human lung cells to polystyrene microplastics significantly retards cell proliferation and triggers morphological changes. *Chemical Research in Toxicology, 34*(4), 1069–1081.

Gouin, T., Avalos, J., Brunning, I., Brzuska, K., de Graaf, J., Kaufmann, J., Koning, T., Meyberg, M., Rettinger, K., Schlatter, H., Thomas, J., van Welie, R., & Wolf, T. (2015). Use of micro-plastic beads in cosmetic products in europe and their estimated emissions to the north sea environment. SOFW, 1–33.

Grand View Research, Inc. (2016). Traffic Road Marking Coating Market Analysis By Product (Paint, Thermoplastic, Preformed Polymer Tape, Epoxy) And Segment - Forecasts To 2022.

Hansen, T. K. (2014). The secondhand clothing market in Africa and its influence on local fashions. *Dresstudy, 64.*

Hernandez, E., Nowack, B., & Mitrano, D. M. (2017). Polyester textiles a source of microplastics from households: A mechanistic study to understand microfiber release during washing. *Environmental Science and Technology, 51*(12).

Hodges, P., & Kelly, P. (2020). Inflammatory bowel disease in Africa: What is the current state of knowledge? *International Health, 12*(3), 222–230.

Iscen, A., Forero-Martinez, N. C., Valsson, O., & Kremer, K. (2021). Acrylic paints: An atomistic view of polymer structure and effects of environmental pollutants. *Journal of Physical Chemistry B, 125*(38), 10854–10865.

Iyare, P. U., Ouki, S. K., & Bond, T. (2020). Microplastics removal in wastewater treatment plants: A critical review. *Environmental Science: Water Research & Technology, 6*, 2664–2675.

Ju, S., Shin, G., Lee, M., Koo, J. M., Jeon, H., OK, Y. S., Hwang, D. S., Hwang, S. Y., Oh, D. X., & Park, J. (2021). Biodegradable Chito-beads replacing non-biodegradable microplastics for cosmetics. *Green Chemistry, 23*, 6953–6963.

Josiane, N., Figoli, A., Weissenbacher, A., Langergraber, N., Marot, B., & Moulin, B. (2013). Wastewater treatment practices in Africa: Experiences from seven countries. *Sustainable Sanitation Practice, 14*, 26–34.

Laitala, K., & Klepp, I. G. (2019). Microfibres from apparel and home textiles: Prospects for including microplastics in environmental sustainability assessment. *Science of the Total Environment, 652*, 483–494. https://doi.org/10.1016/j.scitotenv.2018.10.166

Lant, N. J., Hayward, A. S., Peththawadu, M. M. D., Sheridan, K. J., & Dean, J. R. (2020). Microfiber release from real soiled consumer laundry and the impact of fabric care products and washing conditions. *PLoS ONE, 15*(6), e0233332.

Lassen, C., Foss Hansen, S., Magnusson, K., Noren, F., Bloch Hartmann, N. I., Rehne Jensen, P., Gisel Nielsen, T., & Brinch, A. (2015). Microplastics: Occurrence, effects, and sources of releases to the environment in Denmark (The Danish Environmental Protection Agency).

Mahhun, S. M., & Shams, M. (2022). Acrylic fabrics as a source of microplastics from portable washer and dryer: Impact of washing and drying parameters. *Science of the Total Environment, 834*, 155429.

Mahon, A. M., O'Connell, B. O., Healy, M. G., O'Connor, I. O., Officer, R., Nash, R., & Morrison, L. (2016). Microplastics in sewage sludge: Effects of treatment. *Environmental Science and Technology*, 1–24. DOI: https://doi.org/10.1021/acs.est.6b04048.

Murphy, F., Ewins, C., Carbonnier, F., & Quinn, B. (2016). *Environmental Science and Technology, 50*(11), 5800–5808.

Paruta, P., Pucino, M., & Boucher, J. (2021). Plastics paints and the environment. EA - Environmental Action 2022. Lausanne, Switzerland. ISBN 978-2-8399-3494-7.

Perren, W., Wojtasik, A., & Cai, Q. (2018). Removal of microbeads from wastewater using electrocoagulation. *ACS Omega, 3*, 3357–3364.

PWC. (2015). Real estate building the future of Africa. www.pwc.co.za/realestate.

Romeo, T., Pietro, B., Pedà, C., Consoli, P., Andaloro, F., & Fossi, M. C. (2015). First evidence of the presence of plastic debris in stomach of large pelagic fish in the Mediterranean Sea. *Marine Pollution Bulletin, 95*, 358–361.

Roos, S., & Levenstam, O. (2017). Microplastics shedding from polyester fabrics. Mistra Future Fashion Report. 1. ISBN: 978-91-88695-00-0.

Shah, A. A., et al. (2008). Biological degradation of plastics: A comprehensive review. *Biotechnology Advances, 26*, 246–265.

Shen, M., Zhang, Y., Almatrafi, E., Hu, T., Zhou, C., Song, B., Zeng, Z., & Zeng, G. (2022). Efficient removal of microplastics from wastewater by an electrocoagulation process. *Chemical Engineering Journal, 428*, 131161.

Sillanpaa, M., & Sainio, P. (2017). Release of polyester and cotton fibers from textiles in machine washings. *Environmental Science and Pollution Research, 24*, 19313–19321.

Statista. (2021). Production of polyester fibers worldwide from 1975 to 2020.

Tao, D., Zhang, K., Xu, S., Lin, H., Liu, Y., Kang, J., Yim, T., Giesy, J. P., & Leung, M. Y. K. (2022). Microfibers released into the air from a household tumble dryer. *Environmental Science and Technology Letters, 9*, 120–126.

Tiseo, I. (2021). Global plastics production 1950–2019. Statista.

Turner, A. (2021). Paint particles in the marine environment: An overlooked component of microplastics. *Water rEsearch x, 12*, 100110.

UNICEF/WHO. (2019). Progress on household drinking water sanitation and hygiene 2000-2017: Special focus on inequalities. New York: United Nations Children's Fund and World Health Organization. ISBN: 978-92-415-1623-5.

USAID and East African Trade Investment Hub. (2017). Overview of the second-hand clothing market in East Africa: analysis of determinants and implications.

Watermeyer, G., Epstein, D., Adegoke, O., Kassianides, C., Ojo, O., & Setshedi, M. (2020). Epidemiology of inflammatory bowel disease in sub-Saharan Africa. A review of the current status. *South African Medical Journal 110*(10), 1006–1009.

Webb, P. (2021). Introduction to oceanography. Pressbook. http://rwu.pressbooks.pub/webb.

WHO. (2019). *Microplastics in drinking water*. World Health Organization.

Yan, Z., Liu, Y., Zhang, T., Zhang, F., Ren, H., & Zhang, Y. (2022). Analysis of microplastics in human feces reveals a correlation between fecal microplastics and inflammatory bowel disease status. *Environmental Science and Technology, 56*(1), 414–421.

Zambrano, M. C., Daystar, P. J., Ankeny, M., Cheng, J. J., & Venditti, R. A. (2019). Microfibers generated from the laundering of cotton, rayon, and polyester-based fabrics and their aquatic biodegradation. *Marine Pollution Bulletin, 142*, 394–407.

Zubris, K. A. V., & Richards, B. K. (2015). Synthetic fibers as an indicator of land application of sludge. *Environmental Pollution, 138*(2), 201–211.

Chapter 12
Future Outlook

Abstract The polymer industry impacts almost every aspect of modern life in Africa as well as across the world. Food security, water resource, housing, job creation, health, and environment are among the aspects of society that are connected to the polymer industry. One of the biggest issues the industry currently faces is waste management, in particular, plastic waste. This chapter draws from the discussion in the different chapters of the book to share some insights and future perspectives on selected issues in the plastic and polymer industry in the African region.

Keywords Sustainable urbanization · Crude oil · Sustainable materials · PCR · Policy instrument · Plastic waste

Opportunity for Sustainable Urbanization and The Plastic Industry in Africa

Plastic pollution is becoming a prominent part of discussions on environment, and the environment is currently a key consideration in the risks and opportunities for global development. Many African countries have rapidly become more urbanized with more people living in cities across the world. Since plastics have become essential materials in mordern society, developing a circular plastic economy is a key element to be considered for sustainable urbanization. Previous sections have discussed mechanical and chemical plastics recycling and energy recovery as well as bioplastics production. For sustainable urbanization in the African continent with regard to the plastic industry, a more circular landscape in plastic production and consumption must be adopted. As emerging urbanization, there still exist opportunities to retain some of the elements of the pre-urban era towards a more sustainable urbanization. For example, implementing measures to retain some of the more sustainable materials used in applications which plastics are fast replacing.

A circular plastic economy presents the promise of a sustainable source of plastic products to support this rising urbanization. The plastic industry has the potential to serve as a major source of employment as a circular plastic economy will expand the activities of the industries. A linear plastic economy has its activities limited to sourcing raw materials, production, and supply. A circular plastic economy extends the industry's activity to collection and mining of the millions of tonnes of waste plastics already produced, sorting, cleaning, recycling, and reforming these plastics into new products. These extended activities will engage more diverse participation and skill sets in the industry from the waste pickers to entrepreneurs developing new products and services around the circular plastics industry.

Already we see entrepreneurs across Africa developing a wide range of products and services around recycled plastics. The demand for recycled plastics also seems to be increasing as recent reports show that the price of the most widely recycled plastic PET, has been on the rise since the beginning of 2022 (Evans, 2022). As of March 2022, the global recycled plastic market was valued at 19.5 billion USD and this is expected to rise at a cumulative average growth rate of 8.1% in the next 4 years (Businesswire, 2022). Being able to develop the capacity to cater to this existing demand and a valuable market will significantly aid sustainable urbanization in Africa through developing industries whose growth is tied to meeting the global sustainable development goals.

In its 2016 presentation to the United Nations Consultative Process on Oceans and the Law of the Sea, the World Economic Forum states the development of insights and building economic bases and scientific evidence and communication as important strategies towards achieving a sustainable plastic economy. It also states effective use of technologies of the fourth industrial revolution such as artificial intelligence, machine learning, novel materials, social media, and biological engineering among others is key to achieving a more sustainable plastics economy.

Alternative Raw Materials for Polymers and Crude Oil Demand

While naturally sourced biodegradable plastics and other polymers may still be unable to compete with synthetic ones in some aspects of their applications due to limitations in properties like durability, stability, and mechanical properties. Synthetic plastics are currently mainly sourced from crude oil. However, more sustainable alternatives to fossil fuels are gaining popularity. Manufacturers are already researching and producing plastics like PET and HDPE from alternative sustainable sources such as lignocellulose (Damayanti et al., 2021) and naturally occurring fats and oils like used cooking oil (Vermeiren & Van Gyseghem, 2012).

African countries like Nigeria and Libya and other oil-producing African countries have a large percentage of their GDP attributed to crude oil revenues. For example, 60% of Libya's GDP comes from the oil and gas sector and 69% of its export revenue comes from exports from the oil and gas sector. With its only other main natural resource being gypsum, Libya's economy is largely dependent on its oil and gas sector. In Nigeria, the oil and gas sector represents around 86% of the total export revenue and 10% of the GDP according to OPEC. Nigeria has other non-oil resources such as tin, iron, lead, limestone, and agricultural land among other resources, hence, the relatively lower dependency of its GDP on the oil and gas sector.

Development of alternatives to fossil-derived plastics and polymers inevitably affects the demand for crude oil. With countries like Nigeria, Libya, and Angola having crude oil as their major export, this is expected to impact the economies of the oil-producing countries in Africa. It is important that these countries are able to adapt to this change in demand.

Opportunities in Alternative Polymer Materials

One of the innovations in recent years towards shifting from non-renewable fossil-derived plastics and unsustainable use of wood is the production of moldable wood-like material produced from lignin, natural fibers, and additives (Tecnaro, 2001). This material with the trade name Arboform can be processed just like thermoplastics (Nagele et al., 2002). This gives it the appearance of wood with the processing advantage of plastics. It can be processed into products like furniture, storage boxes, and watches among other products. Arboform can be produced using plastic processing techniques like injection molding and extrusion. This reduces deforestation by reducing the need for natural wood. Instead, waste from processing of wood and other plant materials can be used. Lignin makes up 30% of wood. It is a by-product of paper production following cellulose extraction.

Since lignin is a by-product of paper production, countries with established paper industries are more likely to have access to lignin supply which can be used for the production of products like Arboform. Currently, paper and paperboard production in the African continent as of 2019 is 3,052,491 tonnes according to UNSTAT. In comparison, Asia produced 194,283,122 tonnes in the same year and North America produced 77,629,810 tonnes. These figures suggest that Africa has less access to lignin from paper production when compared to these other regions with producers like China and the United States.

Innovations in Plastic Waste Management

With the recent discovery of plastic-eating microorganisms from which plastic degradation enzymes are extracted (Zhu et al., 2022), some of the potentials include reducing the land area required for landfilling. Conditions for the growth and activities of these microbes can be optimized through well-engineered landfill such that the accumulation rate is reduced by increasing the biodegradation rate of the plastics and polymers in the landfill. Furthermore, the metabolic pathways of these microorganisms can be controlled to obtain useful by-products such as fuel and fertilizers, therefore, providing energy and soil nutrient. In addition to these, harmful impacts of landfilling can be mitigated by controlling the biodegradation process and preventing the production of certain by-products. In order to utilize such innovations, there is a need to effectively manage landfills and install engineered landfills that allow better control of the conditions.

Plastic Pollution is a Global Problem

It should be emphasized here that plastic pollution is a global problem, not just limited to any region. It is estimated that from their first introduction into the global market to the year 2015, around 8.3 billion tons of plastics have been produced (Geyer et al., 2017). At an estimated annual virgin plastic production rate of around 400 million tons, this suggests that, by the end of 2022, another 2800 million tonnes of plastics will have been produced between 2015 and 2022. Thus, it is suggested that the total plastics produced globally by 2022 is around 11.1billion tonnes of plastics. Considering that this annual rate of plastic resin and fibers production is increasing at a compound annual growth rate of 8.4%, these figures are likely to be higher than these estimates.

Despite some countries' increasing recycling rates in recent years, on average, around 79% of the plastic waste ends up accumulating in landfills or other parts of the environment (Geyer et al., 2017). As of 1960, plastics made up less than 1% of municipal solid waste. As in 2005, this has risen to 10% (Jambeck et al, 2015). With the world becoming more globalized and trade routes have become more extensive, plastics produced in one region can reach another region within a day. Also, plastic released into the land is likely to end up in the nearby rivers and other water bodies and eventually in the seas and oceans. From the ocean, the plastics can wash up into beaches in any part of the world and may get washed up inland when it rains or other means. Such plastic pollution anywhere is a problem for the whole world. Taking a regional approach allows us to see what is characteristic of certain regions and the manner in which plastic waste is generated and how strategies best suited to these regions can be applied to stop or reduce the plastic waste generated in the selected region.

Rwanda-Peru Resolution

Rwanda continues its leading role in addressing global plastic pollution in the Rwanda-Peru resolution that was adopted at the United Nations Environmental Assembly 5.2. The resolution which was jointly written by Rwanda and Peru is a crucial step in creating a legally binding treaty that commits nations to take actions to address plastic pollution (UNEP, 2022). The draft resolution is co-sponsored by other member states like Costa Rica, Norway, The European Union, Madagascar, and Senegal. Switzerland, Chile, Uganda, and United Kingdom among others. The UNEA 5.2 conference was held in Nairobi Kenya. Rwanda and Kenya have been actively involved in global environmental issues and both countries have continued to establish themselves as environmentally conscious.

The first step is to set up an intergovernmental negotiating committee whereby representatives of each country come together to agree on legally binding terms in relation to the management of plastics towards a circular plastic economy. The significance of the draft resolution is that it sets the stage and basis from which a global legally binding agreement can be constructed.

One of the significant features of the draft resolution is that it considers not only the management of plastic waste but also looks at the entire life cycle of plastics. Many of the efforts in tackling plastic pollution problem have been from the waste management point of view and legislative actions to ban the use and production of plastics in some countries as other chapters in this book discuss. By taking the approach of looking at the entire life cycle of plastics, this resolution impacts the plastic industry as a whole. In doing so, it includes the plastics in use, the plastics that have been produced, the plastics that are being produced, and those that are yet to be produced. From the raw materials to the post consumer plastics and its return into the plastic cycle.

Much of the actions against plastic pollution have focused on plastics in the ocean. While the significance of the ocean for the planet should not be understated, it is important to note that 80% of the plastics in the ocean originate from plastics that were used and produced on land. Therefore, the resolution in part serves to draw some attention to the need to address plastic on land. The proposed terms listed include those around knowledge and technology sharing towards tackling plastic pollution and the mechanisms through which the implementation of the legal terms will be financed.

Extended Producer Responsibility

Globalization now means that a product can be produced in Spain today and make it to Nigeria within hours. This makes it difficult to track waste generation. Governments can make use of the economic instrument in the form of extended producer responsibility laws which holds the producer and importers responsible for the impact of the production, use, by-products, and end of life of the product on the environment. The producers are therefore not only responsible for the environmental impact of the product during manufacturing but also are responsible for the product throughout the product's life cycle. For example, when manufacturers package products with recyclable plastics, the manufacturer should be responsible for the recovery of the plastics from the user and processing them in an environmentally friendly manner. When the product is shipped to other regions like Africa, regulations and global agreements should ensure that the producer's responsibility extends to the other regions where the product is shipped. Producers can address this by, for example, implementing take-back policies or providing facilities for package refill and reuse. Having such long-term responsibility will motivate producers to consider the environmental impact of the product in all aspects of product design and development.

Data Collection on Plastics and Polymer Production and Waste Generation

Collecting and analyzing data on the production of plastics and polymer products is important for predicting global trends, keeping track of global resources, developing technologies and strategies for addressing plastic and polymer waste management, and much more. For example, data on plastic waste flow has helped identify the leading African rivers through which plastic waste gets to the ocean (Schmidt et al., 2017).

Installing weighbridges in waste dumps/landfills, collecting data on the waste composition, and controlling the access to dumpsites and landfills are among the strategies that contribute to better data collection on waste. Analyzing the composition of waste is important in allowing a multidisciplinary approach to developing technologies and strategies for the recovery of useful materials and energy from waste.

Waste Business and Increasing Demand for PCR

The informal plastic collection has become a relevant industry in some cities in Africa. In Lagos, for example, informal waste collection is carried out by individuals with push carts. These informal collectors collect waste from homes and local businesses for a fee. The fees are often negotiated on the spot. These waste pickers separate valuably reusable, sellable, or recyclable items like plastic bottles from the waste. The rest is then ideally taken to landfills or dumpsites.

Recent years have seen increased demand for PCR (post-consumer recycled) plastics. The labor and logistics account for much of the cost of processing PCR. Africa needs to harness its youthful and populous workforce to take advantage of the current global dynamics in the plastic and polymer industry towards sustainable economic growth. This requires innovative thinking to organize the existing workforce to mine and harness the inherent value in plastic waste that exists in the land and water. In doing so, the plastic and polymer industry can create a cleaner environment and improved sustainable economy.

Microplastics and Potential Impact on Development

Microplastics impact soil quality, hence crop yield and food security, it affects groundwater and aquatic resource conservation. Such that microplastics have emerged as a problem which extends plastic pollution beyond a waste management issue, rather it points out the far-reaching impact of plastics down to the food chain. Understanding that plastics and polymers and other materials that are produced could ultimately end up in our food and even the air we breathe should further emphasize the need for a more circular approach to product design and development.

As discussed in the previous chapter, microplastics tend to remain in water released into the ocean if advanced wastewater treatment is not used. Many countries in the African region have insufficient wastewater treatment facilities/capacity to meet the existing demand. Developing wastewater treatment infrastructure could help reduce the amount of microplastics that gets into the ocean from this region. As emerging urbanization, there is an opportunity to ensure that new technologies being introduced to the market address this newly emerged issue of microplastics. For example, as more urban African cities adopt the use of washing machines and textile washing has been identified as one of the highest sources of microplastic release into the environment (Hernandez et al., 2017), strategies can be implemented to ensure that these washing machines are those fitted with filters or similar technologies to capture and safely dispose of microplastics.

Research on wastewater treatment reveals that the solid residues from wastewater treatment with thermal dehydration, lime treatment, and biological digestion still contain more than 4,000 microparticles/kg of dry weight (Mahon et al., 2016). More advanced wastewater treatment techniques are therefore required for the elimination of microplastics. Advanced methods such as ultrafiltration can successfully remove microplastics from wastewater with high efficiency. However, such infrastructures require high investments. There is therefore a need to develop highly effective yet low-cost wastewater treatment methods that remove microplastics. Methods like electrocoagulation are available low-cost wastewater treatment methods that can effectively remove microplastics (Perren et al., 2018).

Textiles manufacturing can be one that involves less advanced machinery and a significant employer of labor. Countries like Rwanda are looking to the textiles industry as a route to industrializing their economy. It is important that in this new growing industry the problem of microplastics is addressed. Strategies to tackle the issue of microplastics released to the environment include requiring tests for microplastics to be included in regulatory requirements and building more sustainable consumer behavior. Textiles products that have been identified as prone to releasing higher amounts of microplastics can also be avoided. Consumers can also be encouraged to buy better quality textiles that will be used for longer and result in lower microplastic shedding (Sillanpaa & Saino, 2017). Perhaps there needs to be an adjustment on the intention of some African countries like Rwanda to ban the import of used clothing. Rather the ban can be focused on substandard synthetic clothing and clothing which have been identified as more prone to the shedding of microplastics.

Collection and utilization of microplastics can potentially become economically important. Microplastics may potentially make for better efficiency in mechanical and chemical recycling and in enzyme degradation due to increased surface area. Whereas larger plastic pieces may require work input for size reduction to increase surface area during processing, microplastics eliminate this step as they exist in micron size already. This could be an advantage when the technology for capturing and collecting microplastics becomes more efficient on a large scale.

With African countries like Rwanda and Kenya playing the lead roles in setting new policies on global environment management, the laws on the environment in this region are likely to become more sophisticated to accommodate better policies on environmental issues. Furthermore, as the research and technology on microplastics advance and more is known regarding occurrence and management, governments and organizations in Africa and across the world will be better equipped to take the right steps. Taking an example of the current EU policy on sustainable waste management, under this policy, sludge from treatment plants can be applied to agricultural land. Alternatively, the waste can be dried and incinerated or sent to landfills. In view of new information suggesting that some of these sewage sludge may contain microplastics from textiles washing and other sources, such policies, therefore, need to be updated. Since microplastics are nonbiodegradable, anaerobic digestion, composting, or burning do not remove them from the environment. New policies in more sustainable economies should include treatments that are effective in retrieving or converting microplastics. This is important as studies have shown that microplastics in soil can adversely impact crop yield and soil quality. Microplastics have also been detected in groundwater.

References

Businesswire (2022). Global recycled plastic market research report 2022: Industry size, market share of key players, latest developments, and demand forecasts to 2030. *Research and Markets*.

Damayanti, D., Supriyadi, D., Amelia, D., Saputri, R. D., Devi, Y. L. L., Auriyani, W. A., & Wu, H. S. (2021). *Polymers, 13*, 2886.

Evans, J. (2022). Recycled plastic prices double as drinks makers battle for supplies. *The Irish Times*, January 17 2022 at 11. Retrieved April 29, 2022.

Geyer, R., Jambeck, J. R., & Law, K. L. (2017). Production, use and fate of all plastics ever made. *Science Advances, 3*(7), e1700782.

Hernandez, E., Nowack, B., & Mitrano, D. M. (2017). Polyester textiles a source of microplastics from households: A mechanistic study to understand microfiber release during washing. *Environmental Science and Technology, 51*(12).

Jambeck, J. R., Geyer, R., Wilcox, C., Siegler, T. R., Perryman, M., Andrady, A., Narayan, R., & Law, K. L. (2015). Plastic waste inputs from land into the ocean. *Science, 347*, 768–771.

Mahon, A. M., O'Connell, B. O., Healy, M. G., O'Connor, I. O., Officer, R., Nash, R., & Morrison, L. (2016). Microplastics in sewage sludge: Effects of treatment. *Environmental Science and Technology*, 1–24. https://doi.org/10.1021/acs.est.6b04048.

Nagele, H., Pfizer, J., Nagele, E., Inon, E. R., Eistenreich, N., Eckl, W., & Eyere, P. (2002) Arboform ® A thermoplastic, processable material from lignin and natural fibers. In Hu, T. Q. (Ed). *Chemical Modification, properties, and usage of Lignin*. Springer. ISBN 978-1-4615-0643-0.

Perren, W., Wojtasik, A., & Cai, Q. (2018). Removal of microbeads from wastewater using electrocoagulation. *ACS Omega, 3*, 3357–3364.

Schmidt, C., Krauth, T., & Wagner, S. (2017). Export of plastic debris by rivers into the sea. *Environmental Science and Technology, 51*(21), 12246–12253.

Sillanpaa, M., & Sainio, P. (2017). Release of polyester and cotton fibers from textiles in machine washings. *Environmental Science and Pollution Research, 24*(23), 19313–19321.

Tecnaro. (2001). *Composition containing lignins, and/or natural fibers useful in the production of profiled bodies for e.g. automobile industries has a moisture content of 1–20 mass percent*. DE10151386A1. Germany.

UNEP. (2022). *Draft resolution from Rwanda and Peru on an internationally legally binding instrument on plastic pollution 10 Jan 2022*. Available online https://wedocs.unep.org/20.500.11822/3780

Vermeiren, W., & Van Gyseghem, N. (2012). *A process for the production of bi-naphtha from complex mixture of natural occuring fats and oils*. EP2459679A1 European Patent Office.

Zhu, B., Wang, D., & Wei, N. (2022). Enzyme discovery and engineering for sustainable plastic recycling. *Trends in Biotechnology., 40*(1), 22–37.

Chapter 13
Conclusion and Recommendations

Polymers have been in existence even before humans. The polymer industry impacts almost every aspect of life from food, to clothing to shelter and other essential and non-essential aspects. The early trade and industry in Africa produced and used polymeric products such as rubber, silk, and cellulosic fibers, many of which were significant in establishing trade routes between different parts of the continent and between the continent and other parts of the world. The early processing of polymer-based products and many other products mainly made use of man and cattle for skills and work. As the world continues to industrialize and is now in the fourth industrial revolution, more of these processing methods have been replaced by mechanization, automation, synthesis, and artificial intelligence.

Much of the modern polymer products consumed within the African region today depend highly on importation. This can be in part attributed to the fact that the pace of technological advancement hasn't quite caught up with the pace of urbanization and globalization. As African countries continue to urbanize, the consumption trends become more like those of the more industrialized countries. With globalization, goods produced from other parts of the world have become more accessible. This has resulted in a rise in the consumption of synthetic polymers, mainly plastics. The ease of production and seemingly lower financial cost has made these plastics more attractive. However, the cost of this rate of consumption to the environment might be direr.

Plastic pollution and the environment have come to the forefront of discussions in the industry today. Governments of various African countries have used different strategies to address plastic pollution. These include using policy instruments like a plastic ban and implementation of some public waste management infrastructures like public bins and landfills. Rwanda, so far, plays a leading role as the first country to implement a ban on plastics, and for this has earned international recognition as an environmentally conscious country. Yet, there is much more that needs to be done in terms of waste infrastructure and effective implementation of policies on waste in order to tackle plastic waste issues in the continent.

O. Olatunji, *Plastic and Polymer Industry by Region*,
https://doi.org/10.1007/978-981-19-5231-9_13

Entrepreneurs across Africa have successfully made the recycling of plastics and other polymer products like rubber and textiles into profitable businesses. In the informal sectors waste pickers and local crafters have engaged in plastic recycling for profit. Although there is limited data on exactly how much of the waste is being recycled in the continent since the informal sector does a significant amount of recycling, yet there are indications that a significant amount of recycling occurs on small scale. Proper coordination and facilitation of the activities of small-scale waste collection and recycling could potentially help develop more effective data collection and recycling output. Much of the recycling done is mechanical recycling or upcycling of plastics and other polymer products for other applications. More advanced recycling like chemical recycling and energy recovery are yet to be established within the continent on a large scale, although several researchers in academic institutions across the continent have reported works on processes like pyrolysis.

The emergence of the problem of microplastics brings a different dimension to plastic waste management. Microplastics showing up in the food chain, soil, and groundwater is an indication that beyond waste management plastic pollution calls for rethinking and redesigning toxic materials out of polymer products and adopting a more circular economy. Technology for collecting and removing microplastics from the environment is in development. Microplastics in the environment pose a threat to crop yield, water resources, and health. All of which are already being threatened by other factors such as conflict and natural disasters within Africa and globally.

Biodegradable plastics and polymers still have only a small fraction of the plastics market compared to fossil-derived plastics and polymers. While the sophisticated technologies and machinery for large-scale production of advanced bioplastics like PHA may be limited, adopting naturally occurring polymers in applications which over the years have been dominated by fossil-derived plastics can have positive impacts on the reduction of plastic waste, importation of plastics, job creation, revenue generation, health, and environment. For example, replacing the fossil-derived plastic shopping bags with woven shopping bags using natural fibers and already well-established fabric weaving and sewing techniques can have the aforementioned impacts.

With a population of well over a billion, diverse resources which include millions of tonnes of plastic and polymer waste in addition to the existing polymer resources that can potentially be converted to energy and materials, the polymer industry in Africa has a lot of potential to be harnessed. Developing domestic technology, improving technical capacities, proper resource management, and environmental conservation are some of the key considerations that are needed towards developing a sustainable and successful polymer industry in Africa.

Printed in the United States
by Baker & Taylor Publisher Services